I0486964

Mathematical Modeling: An Introduction

William Layton

2022

ii

Contents

Preface

A mathematical model is a black box where numbers come in, calculations are performed and predictions come out. Usually a model contains parameters that must be determined from data for a particular setting before predictions can be made for that specific situation. There are many ways to develop models. The path herein is a classical one where understanding of a phenomena is formalized into equations that have meaning for humans. Predictions made from the resulting equations can then be used to understand the phenomena modelled.

This book was written to give an introduction to interesting and high impact aspects of applied mathematics modeling, calibration and analysis based on a background of basic calculus. The path taken in this introduction is through modeling a system of interest by a collection of ordinary differential or difference equations, calibrating the equations against data, understanding the *qualitative* patterns suggested by the model, checking *quantitative* predictions against actual events and concluding where the next model refinement is needed. Since the goal of modeling is accurate prediction the central modeling questions are

How is the black box, the model, to be constructed?

How to use available data to determine the model parameters?

What are the errors in the model's predictions?

How can the analysis of the model's errors be used to improve future model predictions?

This book's goals are to be casually readable, to introduce a beautiful and interesting subject at an elementary level (one term of differential calculus is enough to get started and two terms are enough to finish) and yet to progress through some mathematically deep ideas, of interest to advanced undergraduates, along the way. The main choices were to begin at the beginning and to stress the geometry. (If you know that when $f'(t) > 0$, $f(t)$ is increasing then that is enough.) Thus, the focus is on models resulting in two ordinary differential equations so the solutions can be plotted in the $x - y$ plane. The model development is geometric with intermediate sketches wherever necessary. These chapters pick topics that are immediately and independently interesting to our life and where mathematics has had a large positive impact on how the challenges of the applications are addressed.

These notes are neither complete nor exhaustive. As an example, proofs are only given when the proof will be understandable by a student reading along and when the idea in the proof has independent value. These notes will be easiest to follow when the reader knows a bit about 2×2 matrices, such as how to represent a system of 2 linear equations in 2 unknowns as $Ax = b$, what the matrix inverse and determinant are and how to calculate the eigenvalues of a 2×2 matrix. Nevertheless, all the material from linear algebra is reviewed in separate sections before use. Thus, inexperience can be replaced by knowledge as linear algebra is needed.

Although all within these pages, including exercises, can be done by paper and pencil, it will be a lot more enjoyable to combine your work in the chapters with a simple program for plotting phase portraits. There are many good ones to choose from and you likely already have one available. For example, mathematics software in packages like MATLAB, Mathematica and Maple all can be used to plot phase portraits. Many phase plane plotters are available around the web. All the phase planes here were plotted only 3 simple ways. The vector fields have been mostly plotted with a computer algebra feature of the program used to type these notes. Phase portraits were either plotted by hand or using the simplest possible applications found by a web search on the term 'phase plane plots'. In Chapter 2 some other possibilities are mentioned. In particular, I highly recommend MATLAB. The exercises of Chapter 2 are also a good places to begin experimenting with one.

Welcome to the world of applied mathematics!

A few words about notation

Dynamic refers to predicting the behavior of some variables, say $x(t)$ and $y(t)$, as a function of time, denoted by t (of course). Given a curve in the $x - y$ plane, it is important to think about it two ways: as a curve given by the collection of points (x, y) satisfying $F(x, y) = 0$ and parametrically as $x = x(t)$, $y = y(t)$ where, as t varies, the curve is traced out. The classic example of such a curve is the unit circle $x^2 + y^2 - 1 = 0$, Figure 1, which is given by the implicit equation

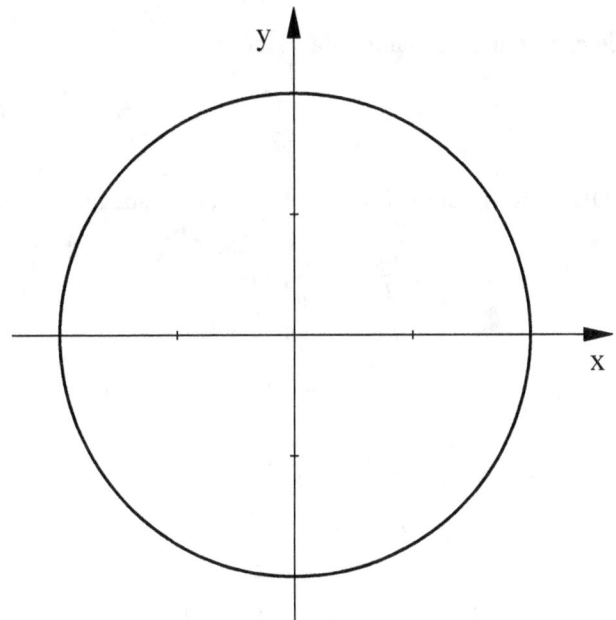

Figure 1: $x(t) = \cos(t), \ y(t) = \sin(t)$
is a circle in the x-y plane

$$F(x, y) := x^2 + y^2 - 1 = 0 \tag{1}$$

and can be written parametrically by

$$\begin{aligned} x(t) &= \cos(t), \\ y(t) &= \sin(t), \end{aligned}$$

$$\text{where } 0 < t < 2\pi.$$

This example of the circle is a good one. For example, since $(\sin(t))' = \cos(t)$ and $(\cos(t))' = -\sin(t)$, it's easy to see that the above $x(t)$, $y(t)$ satisfy the ordinary differential equations (or ODE's):

$$\begin{cases} \frac{dy}{dt} = x, \\ \\ \frac{dx}{dt} = -y \end{cases}$$

and initial conditions

$$\begin{cases} x(0) = 1, \\ \\ y(0) = 0. \end{cases}$$

The chain rule can be used to eliminate t since

$$\frac{dy}{dx} = \frac{\frac{dy}{dt}}{\frac{dx}{dt}}.$$

Dividing the ODE's for y and x (i.e., applying the chain rule) gives:

$$\frac{dy}{dx} = -\frac{x}{y}, \text{ and} \tag{2}$$

$$y = 0 \text{ when } x = 1.$$

Separating variables (also justified by the chain rule) gives

$$ydy = -xdx \Rightarrow \int ydy = -\int xdx, \tag{3}$$

$$\Rightarrow \frac{y^2}{2} + C = -\frac{x^2}{2} + C$$

Thus we arrive back at the implicit equation for the circle $x^2 + y^2 = C$. Since $y = 0$ when $x = 1$, the constant $C = 1$ and the solution to the ODE's can thus be written

$$F(x,y) := x^2 + y^2 - 1 = 0.$$

This is one key idea: given ODEs that $x(t)$, $y(t)$ satisfy, eliminate t and consider an ODE like (0.4) for just y and x and even sometimes an explicit formula $F(x,y) = 0$ for the solution curves $x(t)$, $y(t)$. This reduced picture is called the phase plane of the system. We explore models based on autonomous systems of differential and difference equations allowing this reduction and the analysis that flows from it. A system

$$x' = P(x, y) \text{ and } y' = Q(x, y)$$

is autonomous if, like the above, $P(\cdot, \cdot)$ and $Q(\cdot, \cdot)$ depend only on x and y and not explicitly on t. There is a basic existence and uniqueness theorem from the theory of ODEs that is important for determining the geometry of solution curves.

Theorem 1 *Suppose $P(x, y)$ and $Q(x, y)$ are continuous with continuous partial derivatives[1] ($\frac{\partial P}{\partial x}, \frac{\partial P}{\partial y}, \frac{\partial Q}{\partial x}, \frac{\partial Q}{\partial y}$). Then, for any initial condition*

$$x(0) = x_0, \quad y(0) = y_0$$

the system has a unique solution passing through (x_0, y_0) at $t = 0$. Additionally, since the system is autonomous, given any point (x_0, y_0) there is exactly one solution curve $(x(t), y(t))$ in the $x - y$ phase plane passing through (x_0, y_0).

The first part of the theorem shows that models leading to such systems are well posed: one can analyze the behavior of the model with confidence that there is a well defined solution. The second part of the theorem implies solution curves in the phase plane cannot cross. Curves can be pretty crazy in $3d$ (think of the paths of birds in the sky) but curves in the plane that can never cross can only do a few, universal things. Given this (big) hammer, we now look for nails: problems with available data and for which accurate predictions of the future state have importance to human life.

0.0.1 Acknowledgements

Some figures in early chapters were made by Noel Heitmann and by Astrid Layton in a later chapter. These figures have been a great help to finally finishing this. Noel, Vince Ervin and John Burkardt read versions of this book and gave a lot of useful feedback. Hugo Krispyn allowed me to use his photographs on Rapa Nui. I thank Noel, Astrid, Vince, John and Hugo.

The notes have been developed over 20+ years. During this time I have learned as much from students as from my own teachers. I thank and dedicate this book to both:

to my teachers, and their teachers before them, all the way back to the beginning,
and
to my students and their students after them, all the way to the end.

- William Layton

---

[1] This regularity assumption is improvable.

Chapter 1

Richardson's Model of Arms Races

1.1 Introduction to Richardson's model of arms races

This chapter addresses the questions

What is the true cause of war?

What are the conditions necessary for peace?

It presents a model of arms races due to L. F. Richardson. The model predicts the amount of arms possessed by two blocks of countries at any given time, denoted $x(t)$ and $y(t)$. Next, we introduce a geometric method of understanding the evolution of such a coupled system. This method eliminates time from the system and reduces the problem to one in the x-y plane, called the *phase plane*. The model is tested for the arms race leading to WWI. Its predictive power for that arms race is remarkable.

1.2 Development of the arms race model

"The positive correlations between war preparations and the frequency and intensity of war convinced all authors that preparations provoked war more than they deterred them, thus confirming the arms race theory of war rather than the deterrence theory." - *William Eckhardt*

Tolstoy thought the cause of war to be the pride of leaders. Such an answer leaves us little hope to satisfy the conditions necessary for peace and thus for the future of mankind. L.F. Richardson, 1881-1953, believed initially many wars to be caused by hostility between nations. He later modified this belief to the view that runaway arms races so destabilized relations between countries that small incidents or petty hostilities could ignite a war. It is clear at least that exponential growth of arms budgets, or anything, cannot long continue and must be resolved in one way or another (e.g., war or economic collapse). It is also clear that arms budgets are quantifiable, hence amenable to mathematical study and to testing the error in the model once developed.

One view of many world leaders of the time was that nations build arms as a response to a perceived threat. Unfortunately, increased arms does not reduce the fear of the arms of the opposing country but only increase *their* fear leading to a further build up of their own. This view is well summarized in the following quotes.

Arms races produce feeling of insecurity:

"The increase of armaments, that is intended in each nation to produce consciousness of strength, and a sense of security, does not produce these effects. On the contrary, it produces a consciousness of the strength of other nations and a sense of fear. Fear begets suspicion and distrust and evil imaginings of all sorts. ... The enormous growth of armaments in Europe, the sense of insecurity and fear caused by them, it was these that made war inevitable. ...This is the real and final account of the origin of the Great War."

- Sir Edward Grey (1862 – 1933), British Foreign Secretary, WWI.

"Uncertainty about the intentions of other states is unavoidable, which means that states can never be sure that other states do not have offensive intentions to go along with their offensive capabilities."

- J.J. Mersheimer, in: The Tragedy of Great Power Politics.

Derivation of the model

We assume there are two competing blocks of countries, A and B. The critical variables are obviously their arms expenditures. These are also variables for which reliable and quantitative data is available. Let:

- $x(t) :=$ arms expenditures of block A.

- $y(t) :=$ arms expenditures of block B.

The model of Richardson is based on three assumptions: *mutual fear, limited resources and underlying grievances or goodwill.*

Mutual fear- Arms are built to counter a perceived threat.

"A structural notion in which the self-help attempts of states to look after their security needs tend, regardless of intention, to lead to rising insecurity for others as each interprets its own measures as defensive and measures of others as potentially threatening."

- John Herz.

The assumption of *mutual fear* is that neither country will either voluntarily build up its armaments or unilaterally disarm. Rather, the arms production of each is a response to the amount of arms possessed by its competitor to deter a perceived threat. The following quotes express this idea.

"We must maintain a military force that is capable of deterring any threat to this nation's security,... I wish with all my heart that the expenditures that are necessary to build and to protect our power could be devoted to the programs of peace. But....",

-President L. B. Johnson, January 17, 1968, state of the Union message

"To be prepared for war is one of the most effective means of preserving peace."

–George Washington

"A strong defense is the surest way to peace. Strength makes detente attainable. ... We cannot rely on the forbearance of others to protect this Nation."

–President Gerald R. Ford, address to a joint session of Congress, August 12, 1974.

Assumption 1: Mutual fear. *In the absence of the other, neither country will either voluntarily disarm or build up its armaments. Rather, the arms production of each is a response proportional to the current arms expenditures by the other.*

Limited resources

The second assumption is that arms expenditures cause an economic drag. Arms expenditures must slow as they consume more of the available resources. There are numerous sources supporting this effect.

"Where the army is, prices are high; when prices rise the wealth of the people is exhausted. When wealth is exhausted the peasantry will be afflicted with urgent exactions. With strength thus depleted and wealth consumed the households in the central plains will be utterly impoverished and seven-tenths of their wealth dissipated."

-Sun Tzu, The Art of War (ca. 400 B.C.)

"The problem in defense is how far you can go without destroying from within what you are trying to defend from without."

–Dwight D. Eisenhower.

"For example, defense spending means that the government is pulling away resources from the uses determined by the market and instead using them to buy weapons and supplies and to pay for soldiers and other military personnel. In standard economic models, defense spending is a direct drain on the economy, reducing efficiency, slowing growth and costing jobs."

– Dean Baker, 2009.

Nikolai Leonov, a general in the KGB, described the result of budget increases for the military coming at the expense of investment in the rest of the economy as follows.

"First there was a visible decline in the rate of growth, then its complete stagnation. There was a drawn out, deepening and almost insurmountable crisis in agriculture. It was a frightening and truly terrifying sign of crisis. It was these factors that were crucial in the transition to perestroika."

-Nikolai Leonov

In the runup to WWI both sides were counting on economic drag crippling

the other's economy.

"Believing that there are practically no checks upon German naval expansion except those imposed by the increasing difficulties of getting money, I have had the enclosed report prepared with a view to showing how far those limitations are becoming effective. It is clear that they are becoming terribly effective."

- W. Churchill, then President of the Board of Trade, 1909.

"It is just possible that the effect of convulsively straining her military resources to the uttermost may, by reacting on the economic and social conditions of France, hasten the return of pacific feelings. ..Should the three-year military service entail an income tax, this would also probably have a sobering effect."

- Prince von Bülow, German Chancellor, 1914

The economic drag effect is captured in the following limited resource assumption of the model.

Assumption 2: Limited resources. *The rate of increase of expenditures in any country decreases as expenditures in that country increase.*

Conflicting ambitions and long-standing grievances

A final assumption is that the two blocks have long-standing grievances or goodwill. Grievances could be territorial, economic or other and goodwill is modelled simply as negative grievance. This assumption is motivated by the following report of L.S. Amery.

"With all respect to the memory of an eminent statesman, I believe that statement to be entirely mistaken. The armaments were only the symptoms of the conflict of ambition and ideals, of those nationalist forces, which created the war. ..It was insoluble conflicts of ambitions and not in the armaments themselves that the cause of the war lay."

- L. S. Amery, in reply to Sir Edward Grey in a Parliamentary debate.

Assumption 3: Underlying grievances. *There are permanent underlying grievances between the countries.*

The three assumptions above yield the following system of ordinary differential equations for the expenditures of A and B:

$$\begin{aligned} x'(t) &= ay(t) - mx(t) + r \\ y'(t) &= bx(t) - ny(t) + s \end{aligned} \tag{1.1}$$

where a, b, m, n, r, s are constants, typically positive. We must also know the initial expenditures:

$$x(0) = x_0, \quad y(0) = y_0 \quad \text{(known values)}. \tag{1.2}$$

The equations (1.1) with the initial conditions (1.2) constitute the linear Richardson model of arms races.

To use the model (1.1) to *predict* the future of a specific arms race the parameters a, b, m, n, r and s must be determined. It is certainly easy in principle to assign numbers to these constants if we have data of expenditures. Indeed, if the rate of increase and actual expenditures (x, x'), (y, y') are observed at three different times and inserted into the equations (1.1), then (1.1) reduces to two 3×3 linear systems for the coefficients (a, m, r) and (b, n, s) respectively. Determining values for the coefficients of (1.1) is possible given enough data!

1.3   A Special Case with an Exact Solution

"I consider that I understand an equation when I can predict the properties of its solutions, without actually solving it."
— *Paul A. M. Dirac*

It is worthwhile considering stability of arms races in special cases of (1.1) which are amenable to exact solution. We will use the exact solutions to describe the stability (defined next) of the arms race.

Definition 2 *An arms race $x(t), y(t)$ is **unstable** if, as $t \to \infty$, either $x(t) \to \infty$ or $y(t) \to \infty$. It is **stable** if both $x(t), y(t)$ are bounded for all time. An arms race is **exponentially unstable** if either $x(t) \to \infty$ or $y(t) \to \infty$ at least as fast as $e^{\alpha t}$ for some $\alpha > 0$. An arms race at equilibrium x^*, y^* is **asymptotically stable** if*

$$x(t) \to x^* \text{ and } y(t) \to y^* \text{ as } t \to \infty.$$

To analyze stability of arms races we first consider the case of unlimited budgets and no grievances.
Special Case 1: Unlimited budgets (m = n = 0) and neither grievances nor goodwill (r = s = 0).
In this case (1.1) reduces to

$$x' = ay \text{ and } y' = bx.$$

This can be solved easily by noting that

$$x'' = (ay)' = a(y') = abx, \text{ so } x'' = abx.$$

This is a scalar, linear, constant coefficient, homogeneous ODE whose solution is

$$x(t) = C_1 e^{+\sqrt{ab}t} + C_2 e^{-\sqrt{ab}t}.$$

Thus, the model predicts an exponential unstable arms race with $\alpha = \sqrt{ab}$ when $m = n = 0, r = s = 0$.

In cases where the two equations cannot be reduced to one second order equation, it is still possible to analyze when the model predicts an exponential arms race by calculating the eigenvalues of a 2×2 matrix, explained next.

1.3.1   Eigenvalues of 2×2 Matrices

"Why is eigenvalue like liverwurst?"
  -C. Cullen

Another approach to writing down the exact solution is by using the matrix exponential as follows. In vector notation this takes the form:

$$\overrightarrow{z}' = A\overrightarrow{z}, \quad \text{where } A = \begin{bmatrix} 0 & a \\ b & 0 \end{bmatrix}, \text{ and } \overrightarrow{z}(t) = \begin{bmatrix} x(t) \\ y(t) \end{bmatrix}. \tag{1.3}$$

Recall that e^z is equal to its power series

$$e^z = 1 + z + \frac{z^2}{2!} + \frac{z^3}{3!} + \cdots = \sum_{n=0}^{\infty} \frac{z^n}{n!}.$$

One way to define the matrix exponential is by replacing z by the matrix.

Definition 3 *For A a 2×2 matrix we define e^{At} by*

$$e^{At} := \sum_{n=0}^{\infty} \frac{(At)^n}{n!}.$$

The behavior of the matrix exponential is basic to understanding systems describing growth , decay and oscillation. For example, the solution to $\overrightarrow{z}' = A\overrightarrow{z}$ is formally (and correctly as well)

$$\overrightarrow{z}(t) = e^{At}\overrightarrow{z}(0).$$

The dynamics of the solution of (1.3) $e^{At}\overrightarrow{z}(0)$ are therefore determined by the *eigenvalues* of the 2×2 matrix A.

Definition 4 *If B is the 2×2 matrix:*

$$B = \begin{bmatrix} a & b \\ c & d \end{bmatrix}$$

*then $det(B) := ad - bc$. The **eigenvalues** of the 2×2 matrix B are solutions of the quadratic equation*

$$det(B - \lambda I) = 0.$$

The eigenvalues of a matrix B are traditionally called λ_1 and λ_2. For a general 2×2 matrix

$$B = \begin{bmatrix} a & b \\ c & d \end{bmatrix}$$

we calculate

$$B - \lambda I = \begin{bmatrix} a - \lambda & b \\ c & d - \lambda \end{bmatrix}. \; Thus$$

$$det(B - \lambda I) = 0$$

gives the quadratic equation for $\lambda_{1,2}$

$$(a - \lambda)(d - \lambda) - bc = 0.$$

There are many known properties of eigenvalues of square, $N \times N$, matrices. We next summarize a few for 2×2 matrices. For these each of the following can be proven by a direct calculation.

Proposition 5 *Let a, b, c, d, α be real numbers.*
  a. If A is symmetric, meaning

$$A = \begin{bmatrix} a & b \\ b & d \end{bmatrix},$$

then the eigenvalues of A are real.
  b. If A is skew symmetric, meaning

$$A = \begin{bmatrix} 0 & b \\ -b & 0 \end{bmatrix},$$

then the eigenvalues of A are purely imaginary $\lambda_{1,2} = \pm ib$.
  c. Let $\lambda_{1,2}$ be the eigenvalues of A. Then the eigenvalues of $A^2, \alpha I + A, e^A$ are respectively

$$\lambda_{1,2}^2, \alpha + \lambda_{1,2} \text{ and } e^{\lambda_{1,2}}.$$

  a. If A is upper triangular, meaning

$$A = \begin{bmatrix} a & b \\ 0 & d \end{bmatrix},$$

then the eigenvalues of A are

$$\lambda_1 = a, \lambda_2 = b.$$

  d. For general 2×2 matrices A, B the eigenvalues of $A + B$ are not the sum of the eigenvalues $\lambda(A) + \lambda(B)$

$$\lambda(A + B) \neq \lambda(A) + \lambda(B) \text{ in general.}$$

Example. The eigenvalues of the Richardson matrix A when $m = n = 0, r = s = 0$

$$A = \begin{bmatrix} 0 & a \\ b & 0 \end{bmatrix}$$

are easily calculated. We have

$$det(A - \lambda I) = \lambda^2 - ab = (\lambda - \sqrt{ab})(\lambda + \sqrt{ab}),$$
$$\lambda_{1,2} = \pm\sqrt{ab}.$$

Since one eigenvalue is always negative and one is always positive the *solution will be a combination of exponential growing, $e^{+\sqrt{ab}t}$, and decaying, $e^{-\sqrt{ab}t}$, solutions.* Thus, the eigenvalue calculation agrees with what was concluded above, *in the case of unlimited budgets and neither grievances nor goodwill, the arms race will always be exponentially unstable.*

1.3.2   The Phase Plane

"Complexity grows exponentially with dimension."
  -Anonymous

One fundamental tool for understanding the system (1.1) (and (1.3) as a special case) is the *phase plane* constructed by considering solutions $(x(t), y(t))$ to be curves in the $2d$, $x - y$ plane traced out (parameterized by) t varying. Thus, a $3d$ problem is reduced to $2d$ without losing essential information. These curves are plotted by eliminating time as a variable. Indeed, since

$$\frac{dx}{dt} = ay, \text{ and } \frac{dy}{dt} = bx,$$

it follows that (considering y as a function of x),

$$\frac{dy}{dx} = \frac{\frac{dy}{dt}}{\frac{dx}{dt}} = \frac{bx}{ay}, \text{ or}$$

$$\frac{dy}{dx} = \left(\frac{b}{a}\right)\frac{x}{y}$$

along the curves $x(t)$, $y(t)$. This is an ordinary differential equation for y as a function of x which is easily solvable by separation of variables as follows. Cross multiply and take the antiderivative of both sides. The solution curves $y = y(x)$ in the $x - y$ plane (the phase plane) thus satisfy:

$$\int ay\, dy = \int bx\, dx.$$

Thus we calculate $a\frac{y^2}{2} = b\frac{x^2}{2} + C$ or

$$ay^2 - bx^2 = C, \text{ a constant.} \tag{1.4}$$

The curves (1.4) represent a family of hyperbolas in the $x - y$ plane with asymptotes:

$$\sqrt{a}y \pm \sqrt{b}x = 0.$$

Sketching these curves in the figure below, shows that in this first special case there must be a runaway arms race

$$x(t) \to \infty \text{ and } y(t) \to \infty \text{ as } t \to \infty.$$

One result that follows from the phase portrait is that the ratio of the two countries expenditures approaches a fixed ratio independent of initial conditions, as the curves plotted in Figure 1.1 below reveal.

  Theorem. *If $m = n = r = s = 0$, then, as $t \to \infty$*

$$x(t) \to \infty, \quad y(t) \to \infty$$

$$\text{and}$$

$$\frac{y(t)}{x(t)} \to \sqrt{\frac{b}{a}}.$$

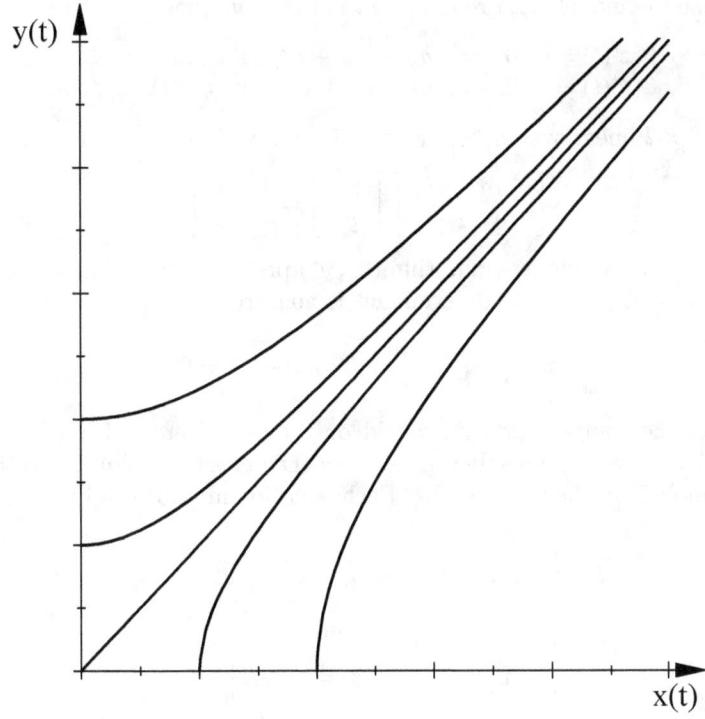

Figure 1.1: Trajectories when $m = n = r = s = 0$ satisfy $x(t) \to \infty, y(t) \to \infty$

1.4 Arms races at equilibrium

"We can do without butter, but, despite all our love of peace, not without arms. One cannot shoot with butter, but with guns."
  - Minister of Propaganda Joseph Goebbels, January 17, 1936.

Since our hope is obviously that runaway arms races are not always inevitable, it is useful to also consider if it is possible to have an arms race at *equilibrium*. An equilibrium arms race corresponds to an equilibrium solution of (1.1). This is a point (x^*, y^*) such that if $x_0 = x^*$ and $y_0 = y^*$ then neither country will alter their armaments.

Definition 6 *An equilibrium (x^*, y^*) is a point such that if $x(0) = x^*$ and $y(0) = y^*$ then $x(t) \equiv x^*$ and $y(t) \equiv y^*$ for all further time $t > 0$.*

If the arms race is at equilibrium then

$$x'(t) = 0 \text{ and } y'(t) = 0.$$

In this case the model (1.1) reduces to the two equations:

$$\begin{aligned}(x'(t) =) \quad 0 &= ay - mx + r, \quad \text{and} \quad x(t) = x_0^*, \\ (y'(t) =) \quad 0 &= bx - ny + s, \quad \text{and} \quad y(t) = y_0^*.\end{aligned} \tag{1.5}$$

This is a 2×2 linear system for the equilibrium arms expenditures (x^*, y^*):

$$\begin{bmatrix} -m & a \\ b & -n \end{bmatrix} \begin{bmatrix} x^* \\ y^* \end{bmatrix} = \begin{bmatrix} -r \\ -s \end{bmatrix}. \tag{1.6}$$

We can solve for a unique equilibrium (x^*, y^*) provided the coefficient matrix of (1.6) is invertible, i.e., if its determinant is nonzero:

$$det \begin{bmatrix} -m & a \\ b & -n \end{bmatrix} = mn - ab \neq 0.$$

Obviously, the solution (x^*, y^*) would only be of interest if x^*, y^* are both nonnegative. Rather than solve this system then check for sign, let us interpret the system (1.5) in the phase space. Each equation in (1.5) is a line in the $x - y$ plane

$$\text{Line } L_1: \qquad y = \frac{m}{a}x - \frac{r}{a}$$

and

$$\text{Line } L_2: \qquad y = \frac{b}{n}x + \frac{s}{n}.$$

The equilibrium (x^*, y^*) is simply the intersection of these two lines, Figure 1.2. Thus, when $r > 0, s > 0$ the system (1.5) can be represented by the following sketch and the condition upon the determinant is simply the condition that the lines not be parallel. Since it's not possible to spend a negative amount on arms, we are also only interested in non-negative values of x^* and y^*. The last two figures show the following.

Theorem 7 *If $r > 0, s > 0$ (grievances), a positive equilibrium of an arms race between two powers can occur if and only if*

$$ab < mn.$$

There are various ways to interpret the condition $ab < mn$. One might say, for example, that $ab < mn$ suggests that the cumulative desire for butter is greater than that for guns.

1.5   A More General Case

"... it occurred to me that ... I would like to spend the first half of my life under the strict discipline of physics, and afterwards to apply that training to researches on living things."
  - L.F. Richardson in: Statistics of Deadly Quarrels

Let us now consider the dynamics of the model (1.1), that is, the arms expenditures $x(t), y(t)$ specifically as time varies.

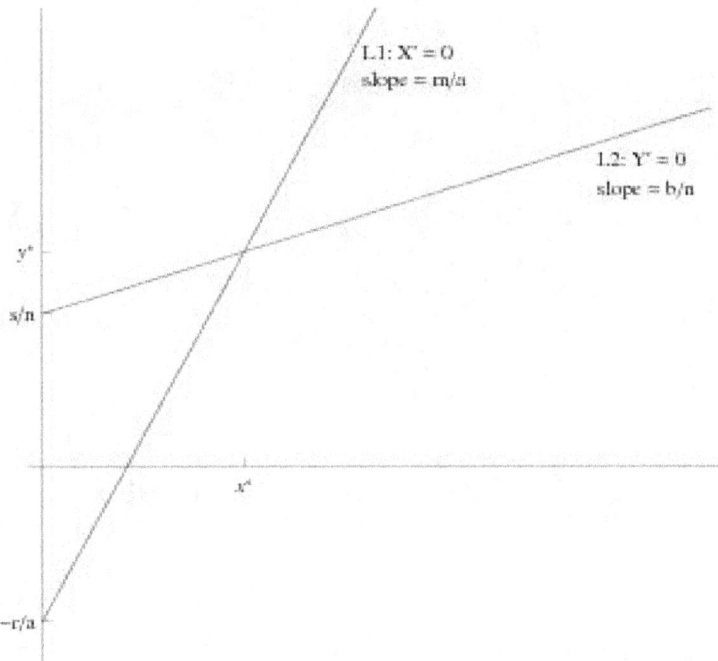

Figure 1.2:
The equilibrium is the intersection of L1 & L2
The lines L1&L2 are called nullclines,
where, resp., $x' = 0$ and $y' = 0$.

Definition 8 *The trajectories of Richardson's model in the $x-y$ phase plane are the parameterized curves $(x(t), y(t)), t > 0$, where $x(t), y(t)$ vary over solutions of (1.1) for different initial conditions $x(0)$, $y(0)$.*

The lines $L1$ and $L2$ in the figures of the last section are important for understanding the behavior of $x(t)$ and $y(t)$. These are called nullclines.

Definition 9 *Consider the system of two ODEs*

$$x' = P(x, y), \quad y' = Q(x, y).$$

The nullclines of the system are the curves where $x' = 0$ and $y' = 0$. Specifically the $x' = 0$ nullcline is the implicitly defined curve $P(x, y) = 0$ and the $y' = 0$ nullcline is the implicitly defines curve $Q(x, y) = 0$.

For Richardson's model the nullclines are thus the lines $L1$ and $L2$.

1.5.1   The case when an equilibrium is possible

The first case of the arms race we consider is when an equilibrium is possible.

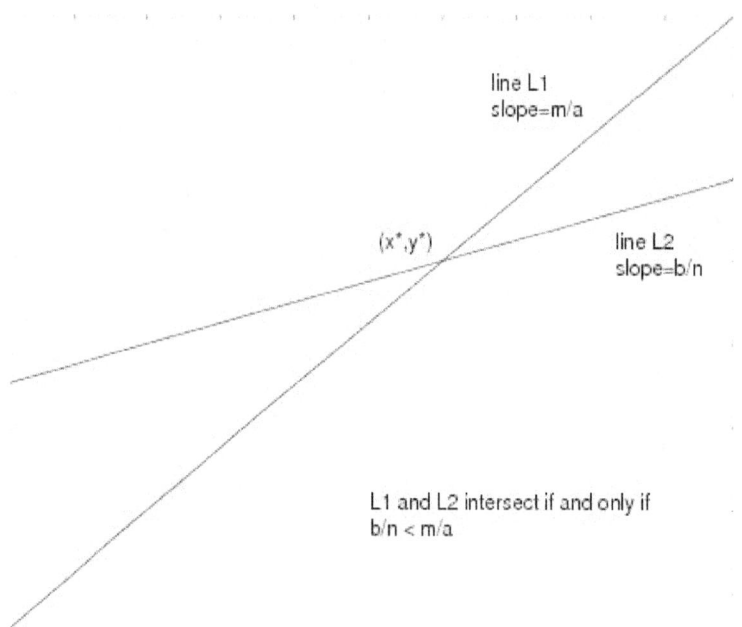

Figure 1.3:   A positive equilibrium, being the intersection of lines L1 & L2, exists if $ab < mn$.

Assumption: $ab < mn$.

Without this assumption, we shall see that no stable arms race is possible. Thus this will be the most important case. Reconsider the nullcline figures of L_1, L_2. The line L_1

$$y = \frac{m}{a}x - \frac{r}{a}$$

is the curve where $x'(t) = 0$. (Such curves are called *nullclines*). Thus $x'(t)$ must be positive on one side of the line L_1 and negative on the other. Where $x'(t) > 0$, we know $x(t)$ is increasing ($x(t)$ moving to the right) and similarly decreasing ($x(t)$ moving to the left) where $x'(t) < 0$. Thus, $x(t)$ must increase on one side of L_1 and decrease on the other. Similarly, $y(t)$ must increase on one side of L_2

$$y = \frac{b}{n}x + \frac{s}{n}$$

and decrease on the other. Testing a few points determines which side is which.

To be specific, consider the nullclineL_1. On L_1, $x' = 0$ so trajectories must cross L_1 vertically. It is easy to check (e.g., by inserting $(x, y) = (0, 0)$) that

- Above L_1,

$$x' = ay - mx + r > 0.$$

Thus $x(t)$ is increasing above L_1.

• Below L_1,
$$x' = ay - mx + r < 0.$$

Thus $x(t)$ is decreasing below L_1.

The division of the plane into regions where $x(t)$ is increasing and decreasing (where trajectories more to the right and left respectively) is depicted in the next figure. A horizontal arrow $\rightarrow$ means that $x(t)$ is increasing and $\leftarrow$ means $x(t)$ is decreasing, Figure 1.4.

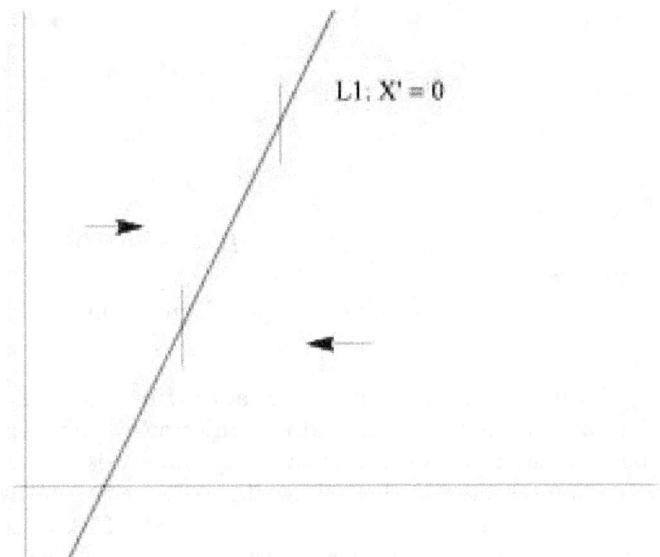

Figure 1.4:
$$\begin{array}{l} x' < 0 \text{ to the left of L1,} \\ x' = 0 \text{ on L1 and} \\ x' > 0 \text{ to the right of L1.} \end{array}$$

Similarly, checking the sign of y' shows that:

• Above L_2,
$$y' < 0$$
so $y(t)$ is decreasing there.

• Below L_2,
$$y' > 0$$
so $y(t)$ is increasing there.

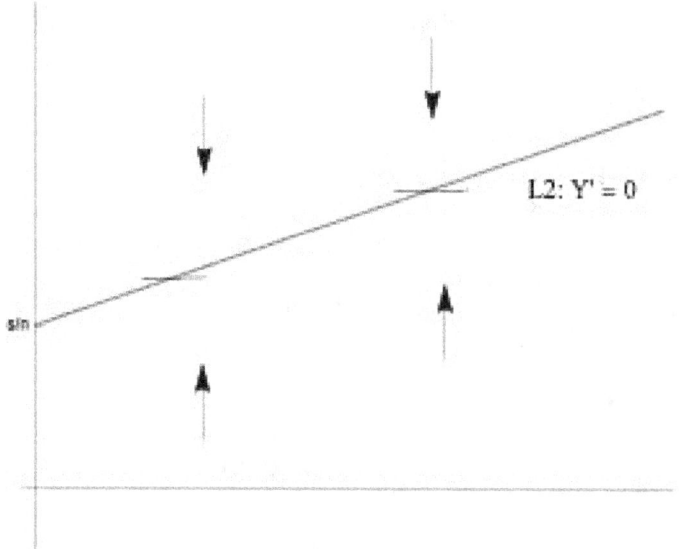

Figure 1.5: Above L2 $y' < 0$ ($y(t)$ is decreasing),
            on L2 $y' = 0$ and
            below L2 $y' > 0$ ($y(t)$ is increasing).

The $y'(t)$ information is represented schematically in Figure 1.5. A vertical arrow like ↑ means $y(t)$ is increasing and the trajectories are moving upwards and analogously ↓ means trajectories are moving downwards.

The information depicted in the last two figures on when $x(t), y(t)$ are increasing or decreasing is combined into Figure 1.6 next. The trajectory directions depicted above suggest that the trajectories $(x(t), y(t))$ must approach the equilibrium point (x^*, y^*) as $t \to \infty$. This can be checked numerically for specific numerical values of a, b, m, n, r, s. There are many convenient programs for plotting phase portraits. Indeed, Figure 1.7 next presents some examples of trajectories directions and a few solutions to the Richardson model.

This geometric argument is already a proof that the arms race is self-stabilizing when $ab < mn$. It is also possible to give an analytical proof using a Lyapunov function, an idea from the theory of ordinary differential equations. This analysis uses the following result from algebra[1].

Proposition 10 *The quadratic form*

$$Ax^2 + Bxy + Cy^2$$

is negative for non-zero x, y if $A < 0, C < 0$ and

$$B^2 - AC < 0.$$

---

[1] This is a result that we all see in high school algebra but forget because we never use it!

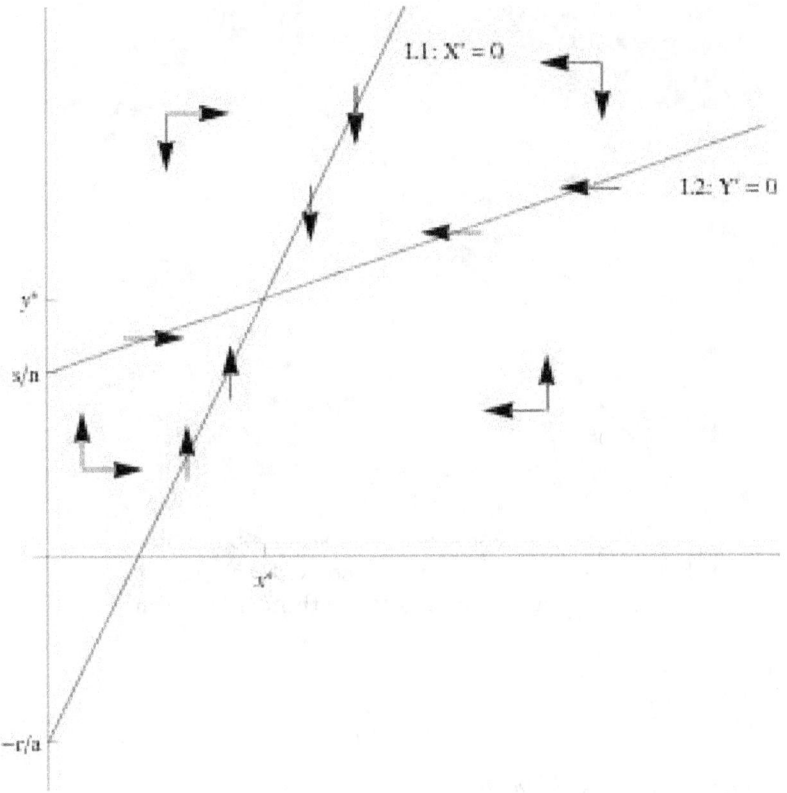

Figure 1.6:   Combining the last two figures gives equilibria, nullclines and trajectory directions.

For example

$$-2x^2 + Bxy - 8y^2 \leq 0$$

for

$$B^2 < AC = 16 \text{ or } -4 < B < 4.$$

There is a linear algebra interpretation of this algebra result because

$$Ax^2 + Bxy + Cy^2 = (\; x \quad y \;) \begin{bmatrix} A & \frac{1}{2}B \\ \frac{1}{2}B & C \end{bmatrix} \begin{pmatrix} y \\ y \end{pmatrix}.$$

This is non-negative provided the eigenvalues of the above 2×2 matrix are non-negative.

Using this algebra result, we now prove stability when $ab < mn$.

Theorem 11 *Suppose a, b, m, n, r and s are all positive and*

$$ab < mn.$$

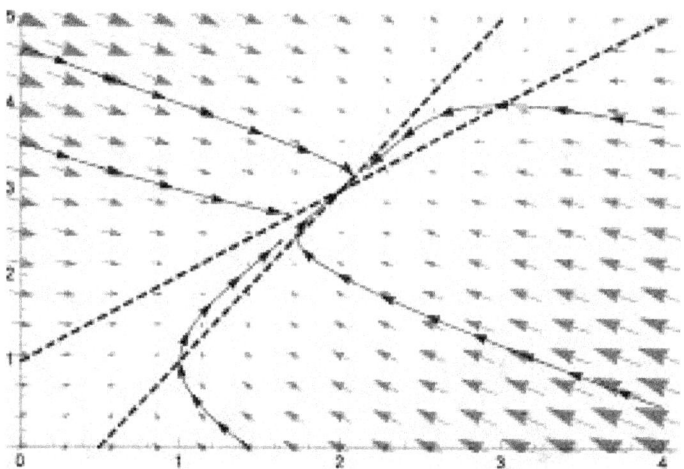

Figure 1.7:   For specific values of a,b,m,n,r,s
                   exact trajectory directions
              are easily plotted with many programs.

Then, as $t \to \infty$,

$$x(t) \to x^* \ and \ y(t) \to y^*$$

and any arms race is self-stabilizing.

Proof. The following steps in the proof are a commonly strategy for stability proofs in ODE's.
Step 1: We derive an equation for the distance of the trajectory $(x(t), y(t))$ to the equilibrium point (x^*, y^*). Indeed, let

$$\begin{aligned} d_1(t) &= x(t) - x^*, \\ d_2(t) &= y(t) - y^*. \end{aligned}$$

By subtraction, d_1 and d_2 satisfy the differential equations:

$$d_1' = ad_2 - md_1,$$

$$d_2' = bd_1 - nd_2, \tag{1.7}$$

which is the same as the Richardson model with $r = 0$ and $s = 0$.
Step 2: Construct

$$V(t) := \alpha d_1(t)^2 + \beta d_2(t)^2$$

with α, β positive parameters. Thus, $V(t) \to 0$ if and only if $(x(t), y(t))$ approaches the equilibrium point (x^*, y^*). Pick the free parameters α and β so that $V(t)$ decreases to zero (by showing $V'(t) < 0$). Indeed, we calculate V' using the chain rule,

$$
\begin{aligned}
V'(t) \quad &= \quad \begin{aligned}[t] &2\alpha d_1 d_1' + 2\beta d_2 d_2' = \\ &\text{using (1.7) for } d_{1,2}' \end{aligned} \\
&= \quad 2\alpha d_1[ad_2 - md_1] + 2\beta d_2[bd_1 - nd_2] \\
&= \quad -(2am)d_1^2 + (2\alpha a + 2\beta b)d_1 d_2 - (2\beta n)d_2^2.
\end{aligned} \tag{1.8}
$$

Step 3: Pick $\alpha > 0$ and $\beta > 0$ so that the right hand side of (1.8) is negative when d_1, d_2 are nonzero.

This is an algebraic calculation since the right hand side of (1.8) is a quadratic form in two variables d_1, d_2. Indeed, from algebra we know that the right hand side of (1.8) is negative provided the discriminant of the quadratic form is negative. In other words, if

$$
(2\alpha a + 2\beta b)^2 - 4(2\alpha m)(2\beta n) < 0. \tag{1.9}
$$

All we need to show is that there exists positive parameters α and β such that this holds. Often, picking α and β carefully will make the analysis easier. Here, it's a piece of (algebraic) cake. We pick

$$
\alpha = \frac{1}{2}b \text{ and } \beta = \frac{1}{2}a
$$

then (1.9) simplifies to:

$$
(2ab)^2 < 4(mb)(na). \tag{1.10}
$$

Simplifying gives that this holds if and only if $ab < mn$, which is exactly the condition we suppose.

Thus, $V'(t) < 0$ and $V(t)$ is decreasing. By being a bit more careful we can even show

$$
V'(t) \leq -\alpha V(t)
$$

for some positive α. Then,

$$
V(t) \leq e^{-\alpha t}V(0)
$$

so $V(t) \to 0$ exponentially fast. Thus, as $t \to \infty$,

$$
V(t) \to 0
$$

and thus

$$
x(t) \to x^* \text{ and } y(t) \to y^*.
$$

∎

The function $V(t) := \alpha(x(t) - x^*)^2 + \beta(y(t) - y^*)^2$ used in the proof is called a **Lyapunov function**. For $\alpha > 0, \beta > 0$ it has the three properties that:

- $V(t) \geq 0$;

- $V(t) = 0$ if and only if $x(t) = x^*$ and $y(t) = y^*$;

- $\frac{d}{dt}V(t) < 0$ along non-equilibrium solutions $x(t), y(t)$ of the system.

1.5.2 The case when there is no positive equilibrium

"Every gun that is made, every warship launched, every rocket fired signifies, in the final sense, a theft from those who hunger and are not fed, those who are cold and are not clothed. This world in arms is not spending money alone. It is spending the sweat of its laborers, the genius of its scientists, the hopes of its children. The cost of one modern heavy bomber is this: a modern brick school in more than 30 cities. It is two electric power plants, each serving a town of 60,000 population. It is two fine, fully equipped hospitals. It is some fifty miles of concrete pavement. We pay for a single fighter with a half-million bushels of wheat. We pay for a single destroyer with new homes that could have housed more than 8,000 people. . . . This is not a way of life at all, in any true sense. Under the cloud of threatening war, it is humanity hanging from a cross of iron."

-President D.D. Eisenhower, 1953.

When fear makes production of weapons more highly valued than economic activity (i.e., guns valued more highly than butter)

$$ab > mn$$

there is no equilibrium. The previous analysis of the trajectory directions can all be repeated, leading to Figure 1.8 below.

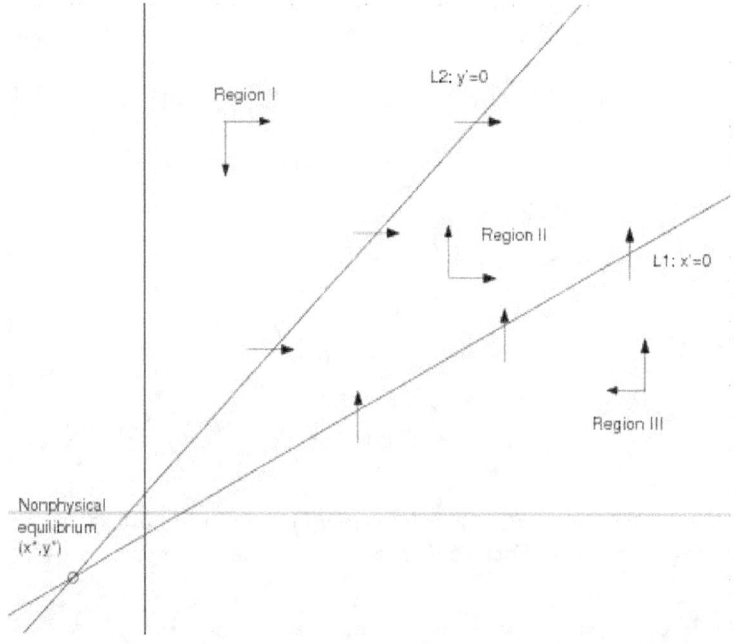

Figure 1.8: When $ab > mn$, runaway arms race is inevitable

If the initial expenditures are in region II then

$$x(t) \to \infty \text{ and } y(t) \to \infty.$$

If the initial expenditures are in region I (region III) then $y(t)$ (respectively $x(t)$) will decrease and the other will increase until the trajectory crosses into region II, whereupon both grow without bound. In all cases, *a runaway arms race is unavoidable when $ab > mn$* as shown in the last figure.

Theorem 12 *Let a, b, m, n, r, s be positive. If*

$$ab > mn$$

any initial condition must result in an unstable arms race: $x(t) \to \infty$ and $y(t) \to \infty$.

1.5.3   The Effects of Goodwill

Trade, internal or international, is the harbinger of goodwill among men, and peace on earth. The opposite of trade is isolation, and isolation is a mark of decadence, of a return to a caveman economy.
-Frank Chodorov (1887-1966)
Goodwill was interpreted by Richardson as persistent *negative hostility*,

$$\text{goodwill} \Leftrightarrow r < 0 \text{ and } s < 0,$$

and reflected in international trade between nations. Since wars disrupt such trade, goodwill intuitively should act as a brake on arms races. The cases of negative r and s can be analyzed following the steps in the previous cases. There are two interesting cases:

$$ab < mn \quad \text{and} \quad ab > mn.$$

Case 1: ab < mn and goodwill.

When $ab < mn$ the nullclinesL_1 and L_2 and trajectory directions divide the positive quadrant into three regions. Clearly, for any initial condition a trajectory must first enter the middle region II, thereupon both countries armaments decrease until one or both have disarmed, Figure 1.9.
Case 2: ab > mn and goodwill.

The trajectory directions and nullclines for the case $ab > mn$ and $r > 0, s > 0$ are sketched in Figure 1.10. Figure 1.10 shows that the outcome of the arms race will depend upon the initial amount of armaments that each country possesses. Indeed:

- If $(x(0), y(0)) \in$ *Region I* then a runaway arms race occurs.

- If $(x(0), y(0)) \in$ *Region III* then one country will disarm.

- If $(x(0), y(0)) \in$ *Region II or IV* then trajectories will exit into either I or III and the outcome will depend upon whether the trajectory crosses into I or III first.

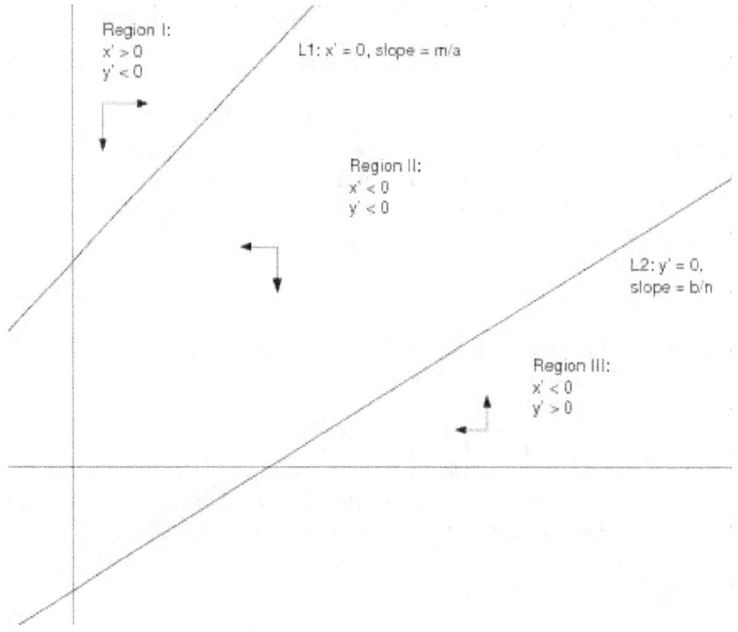

Figure 1.9:   If $ab < mn$ and goodwill, one country will disarm: Trajectories enter Region II. Then $x \to 0$ or $y \to 0$.

1.6   Parameter Determination for the Model

"Prediction is hard, especially about the future."
  -attributed to many people including Niels Bohr

A *mathematical model* is a black box where numbers go in and predictions come out. The goal of a model is thus to make predictions that can be checked against actual events. Thus, the variables used must be ones for which data is available and the model parameters must be available. The model solution is then calculated, usually by numerical methods, and model's prediction is compared against the actual values of the variables modelled. In Richardson's model the variables are arms budgets but the parameters, such as the "mutual fear coefficient", are not ones that can be looked up as easily as the density of water for example. Thus, one issue Richardson confronted is to use the available data for arms budgets to determine coefficient values. This is one example of a general mathematical procedure called *parameter identification*.

Richardson's model is

$$x'(t) = ay(t) - mx(t) + r, \ \text{with} \ x(0) \ \text{given},$$
$$y'(t) = bx(t) - ny(t) + s, \ \text{with} \ y(0) \ \text{given}.$$
$$(1.11)$$

The model contains 8 parameters that must be specified before it can be used to predict future arms budgets. One cannot look up the "mutual fear coefficient"

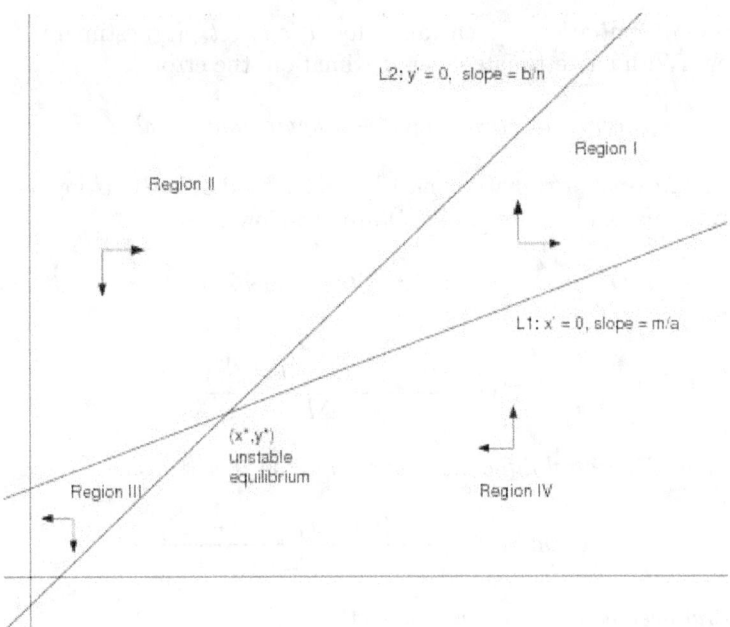

Figure 1.10:   Trajectory directions when $ab > mn$ and goodwill: Both disarmament and unstable race are possible.

for each block in the same way one looks up the density of water. Thus, the coefficient values must be inferred from arms budgets. For the arms race leading to WWI, the following budgets are summarized from p. 32 of Richardson, *Arms and Insecurity* for the countries in each block.

Country\Year	1909	1910	1911	1912	1913
France	48.6	50.9	57.1	63.2	74.7
Russia	66.7	68.5	70.7	81.8	92.9
Germany	63.1	62.0	62.0	68.2	95.4
Austria-Hungary	20.8	23.4	23.4	25.5	26.9

Arms Budgets in Millions of Pounds

Richardson noticed that if $x(t), y(t), x'(t), y'(t)$ are all known then the model becomes a linear system[2] for the unknown parameters $a, b, m, n, r, s, x(0), y(0)$.

Thus for parameter determination for the model the issue became one of

---

[2] Already interesting mathematics is needed. If there are 8 data points with all x, y, x', y' known, then it becomes 8 equations for 8 variables. In this case one simply solves an 8×8 square linear system for the coefficients. If there are more than 8 data points, the coefficients are found by doing a *linear least squares fit* (also known as a *linear regression*). If there are fewer then 8 then the system is undetermined and becomes a problem in compressed sensing for which other mathematical algorithms are applied.

numerical differentiation: Given values for $(t, x(t)), (t, y(t))$ estimate $x'(t), y'(t)$ accurately. With any estimate or approximation, the error

$$error := \ true \ value \ - \ approximate \ value$$

in the estimate or approximation must be understood before trusting its validity.

Basic estimators for $x'(t)$ and $x(t)$ are as follows.

Definition 13 *Let the time step* $\triangle t > 0$ *be given. The central difference approximation to* $x'(t)$ *is*

$$x'(t) \simeq \frac{x(t + \triangle t) - x(t - \triangle t)}{2 \triangle t}.$$

The error in the central difference approximation to $x'(t)$ *is*

$$error := x'(t) - \frac{x(t + \triangle t) - x(t - \triangle t)}{2 \triangle t}.$$

The central average approximation to $x(t)$ *is*

$$x(t) \simeq \frac{x(t + \triangle t) + x(t - \triangle t)}{2}.$$

The error in the central average approximation to $x(t)$ *is*

$$error := x(t) - \frac{x(t + \triangle t) + x(t - \triangle t)}{2}.$$

We have the following error result on the above two approximations.

Theorem. *The error in the central difference approximation to* $x'(t)$ *is*

$$error := x'(t) - \frac{x(t + \triangle t) - x(t - \triangle t)}{2 \triangle t}.$$

The error satisfies

$$x'(t) - \frac{x(t + \triangle t) - x(t - \triangle t)}{2 \triangle t} = -\frac{1}{6} \triangle t^2 x'''(\xi) \ for \ some \ \xi, t - \triangle t < \xi < t + \triangle t.$$

The error in the central average approximation to $x(t)$ *is*

$$error := x(t) - \frac{x(t + \triangle t) + x(t - \triangle t)}{2}.$$

The error satisfies

$$x(t) - \frac{x(t + \triangle t) + x(t - \triangle t)}{2} = -\frac{1}{2} \triangle t^2 x''(\xi) \ for \ some \ \xi, t - \triangle t < \xi < t + \triangle t.$$

proof: Consider first the central difference approximation to $x'(t)$. Let $e(t)$ denote the error in this approximation to $x'(t)$. *Expand $e(t)$ in a Taylor series about t. This gives*

$$
\begin{aligned}
e(t) &= x'(t) - \frac{1}{2\triangle t}\left\{x(t+\triangle t) - x(t-\triangle t)\right\} \\
&= x'(t) - \frac{1}{2\triangle t}\left\{ \begin{array}{l} \left(x(t) + \triangle t x'(t) + \frac{1}{2}\triangle t^2 x''(t) + \frac{1}{3!}\triangle t^3 x'''(\xi)\right) \\ - \left(x(t) - \triangle t x'(t) + \frac{1}{2}\triangle t^2 x''(t) - \frac{1}{3!}\triangle t^3 x'''(\xi)\right) \end{array} \right\} \\
&\quad \textit{(after algebraic simplification)} \\
&= -\frac{1}{6}\triangle t^2 x'''(\xi) \textit{ for some } \xi, t - \triangle t < \xi < t + \triangle t.
\end{aligned}
$$

The error analysis of the central average approximation follows similarly.

The theorem asserts that, looking at arms expenditures for France in 1909 and 1910 for example,

$$\frac{50.9 - 48.6}{1}$$

is a *reasonable estimate* of the rate of increase of France's arms expenditures in 1909.5 (June 1909). If $\triangle t$ (here 0.5 years) were smaller the difference approximation above would be a *good approximation*.

To identify parameters and test the model Richardson used data from the arms budgets of the competing blocks preceding World War I. In the main, these competing blocks consisted of France and Russia against Germany and Austria-Hungary. He supposed that

Assumption for WWI: *The mutual fear and economic drag coefficients for each block were comparable in size.*

Equating them led to a system with only four coefficients to be determined, a, m, r and s:

$$
\begin{aligned}
x' &= ay - mx + r \\
y' &= ax - my + s.
\end{aligned}
\tag{1.12}
$$

This system has the property that the sum of the total arms expenditures of *all* sides can be easily predicted. Indeed, set

$$
\begin{aligned}
z(t) &= x(t) + y(t) \\
&= \text{total arms expenditures of all combatants.}
\end{aligned}
$$

Taking the above arms budgets, adding (to get total spent $= Z(t)$) then taking differences gave Richardson the following table of data (in which the units are millions of pounds sterling) from his book Arms and Insecurity, page 32 of the

1960 edition.

Year t	Total Arms Budgets Z(t)	Increases Z'(t) $\approx$ Z(t) − Z(t − 1)
1909	199.2	
		5.6
1910	204.8	
		16.1
1911	214.9	
		23.8
1912	238.7	
		50.3
1913	289.0	

The increases, $z'(t)$, are in between years while the expenditures are on years. To make the time consistent, we calculate mid-year averages. Thus "1909.5" means the averages of 1909 and 1910, representing June of 1909. This data is:

t	1909.5	1910.5	1911.5	1912.5
$Z(t)$	202.0	209.8	226.8	263.8
$Z'(t)$	5.6	10.1	23.8	50.3

data: $Z,\ Z'$

The question now is how to use this data to determine model parameters.

Adding the two equations in (1.14) gives a closed equation for $z(t)$ with only two unknown parameters $a - m$ and $r + s$:

$$z'(t) = (a - m)z(t) + (r + s) = (a - m)[z(t) + (r + s)/(a - m)]. \qquad (1.13)$$

This equation is a line in the (z', z) plane with slope $(a - m)$ and vertical intercept $(r + s)$. The data points are given in Figure ??.

Since each data point has error, the correct line is found by a linear least squares fit. In statistics this is called a linear regression and most scientific calculators will do all the calculations needed to produce the line of best fit. Richardson produced the parameter values from the slope and intercepts of the linear least squares fit of the data points.

He obtained:

$$m(= n) \doteq 0.2,\ a(= b) \doteq 0.9 \text{ and } r + s = -136.5$$

1.7   Testing the Model

"An nescis, mi fili, quantilla prudentia mundis regitur?" (know you not, my son, with how little wisdom this world is governed?)
  - Count Axel (Gustafsson) Oxenstierna (1583 – 1654), Swedish statesman.

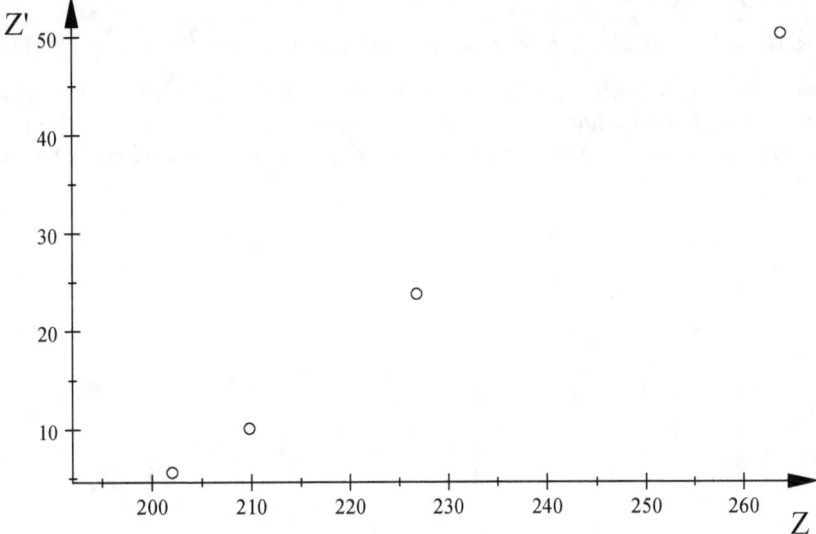

Figure 1.11: (Z, Z') data

"The positive correlations between war preparations and the frequency and intensity of war convinced all authors that preparations provoked war more than they deterred them, thus confirming the arms race theory of war rather than the deterrence theory."

- William Eckhardt

To test the model Richardson used data from the arms budgets of the competing blocks preceding World War I. In the main, these competing blocks consisted of France and Russia against Germany and Austria-Hungary. He supposed that

Assumption for WWI: *The mutual fear and economic drag coefficients for each block were comparable in size.*

Equating them led to a system with only four coefficients to be determined, a, m, r and s:

$$\begin{aligned} x' &= ay - mx + r \\ y' &= ax - my + s. \end{aligned} \qquad (1.14)$$

This system has the property that the sum of the total arms expenditures of *all* sides can be easily predicted. Indeed, set

$$\begin{aligned} z(t) &= x(t) + y(t) \\ &= \text{total arms expenditures of all combatants.} \end{aligned}$$

Adding the two equations in (1.14) gives a closed equation for $z(t)$ with only two unknown parameters $a - m$ and $r + s$:

$$z'(t) = (a - m)z(t) + (r + s) = (a - m)[z(t) + (r + s)/(a - m)].   \quad (1.15)$$

This equation can easily be solved by using an integrating factor. However, it is easier to understand what it is telling us by noting that it is a line in the (z', z) plane with slope $(a - m)$ and vertical intercept $(r + s)$, as sketched in the next figure.

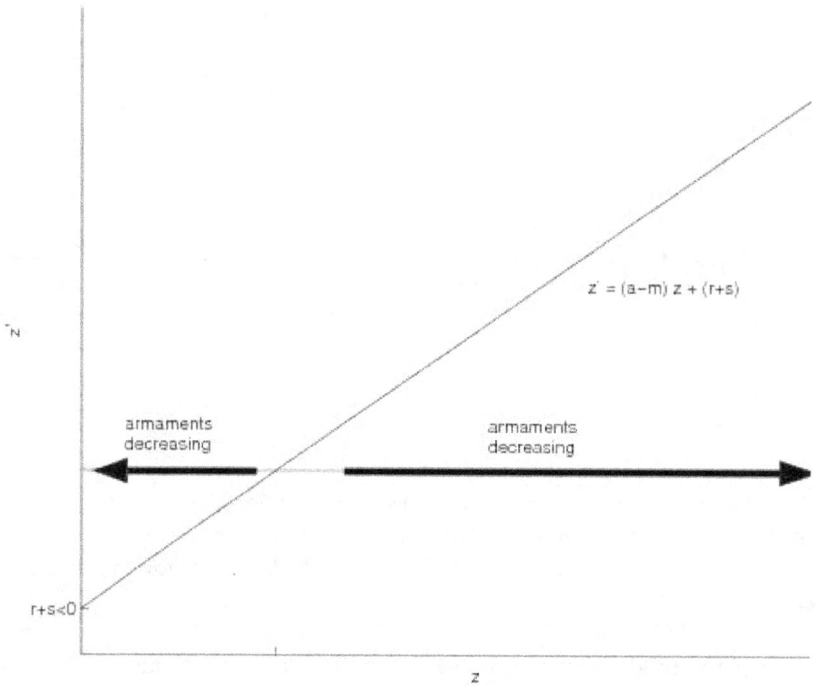

Figure 1.12: The z, z' line

Thus, one qualitative test of the models descriptive powers is to simply check if the real expenditures follow this linear pattern. It is remarkable that the real arms expenditures do indeed point to a linear relationship between z and z'. Using a linear least squares fit, Richardson produced the values

$$m(= n) \doteq 0.2, \ a(= b) \doteq 0.9 \text{ and } r + s = -136.5$$

from the slope and intercepts of the linear fit. The data points and the linear least squares fit are plotted below. With his values $m(= n) \doteq 0.2$, $a(= b) \doteq 0.9$ and $r + s = -136.5$ the product

$$ab = 0.81 \text{ and } mn = 0.4$$

and thus

$$ab > mn.$$

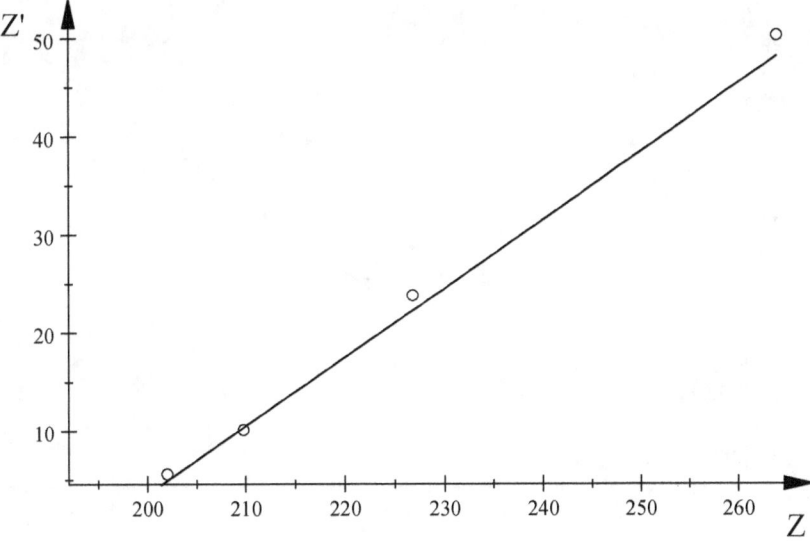

Figure 1.13: (Z, Z') data & Richardson's fit

Since there was *goodwill* in the form of lucrative trade, travel and personal contacts between the nations involved, we are precisely in the case when both war and peace are possible depending on initial arms budgets. If the initial budgets are too large then an unstable arms race must occur. If the expenditures are below a critical level, tensions between the blocks will dissipate and the expenditures of both blocks eventually decrease. The critical line in the $z - z'$ plane and the associated data points are depicted in the next figure.

From this figure, the critical spending level is 195 million. If the combined expenditures had been less than 195 million, then $z'(t) < 0$ and no arms race would have occurred. In 1909 the actual expenditures were $199.2\,(> 195)$ and Europe was caught in an arms race which spiraled out of control and exploded into a world war! As Richardson wrote in his 1960 book *Arms and Insecurity*:

"As love covereth a multitude of sins, so the goodwill between opposing alliances would just have covered 194 million pounds of defense expenditure on the part of the four nations concerned. The actual expenditure in 1909 was 199 million; and so began an arms race which led to World War I."

1.8　References for Chapter 1

The modeling process is: identify variables where data is available, determine how the current changes are related to the current state of the system, develop properties of the model, calibrate the model with data and compare model

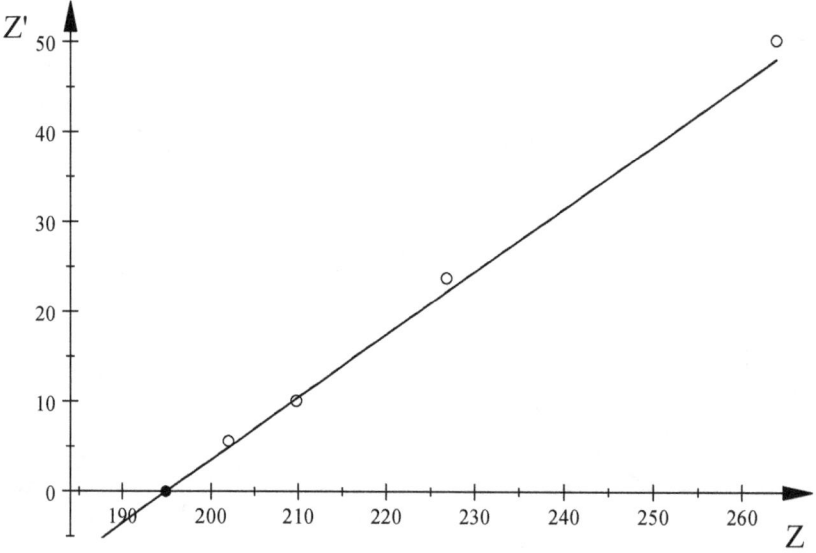

Figure 1.14: Total arms expenditures preWWI: almost perfect linear fit.

predictions to actual events. We have seen the value of this modeling process for understanding when seemingly complex phenomena are driven by simple interactions. We have also seen that when a model can be simplified to two components (two variables) the phase plane gives a powerful tool for understanding model behavior. This process was developed by Richardson for arms races into models that still have relevance today. We hope that the mechanistic operation of the arms race leading to WWI is no longer active. Nevertheless, as a final remark, we offer the following observation of current events.

"In many ways, the situation in Asia today closely resembles that obtaining in Europe in the years preceding the First World War."

- from: http://www.vifindia.org

L.F. Richardson, Mathematics of War and Foreign Politics, pages 1240-1253 in: The World of Mathematics, vol. IV (J. R. Newman, editor) , Simon and Schuster, New York, 1956.

L.F. Richardson, Generalized Foreign Policy, British J. of Psychological Monographs, Supplement 23, 1939.

L.F. Richardson, Arms and Insecurity: A Mathematical Study of the Causes and Origins of War, Boxwood Press, Pittsburgh, 1960.

T.T. Saaty, Mathematical Models of Arms Control and Disarmament, Wiley, New York, 1968.

M. Olinick, An Introduction to Mathematical Models in the Social and Life Sciences, Addison-Wesley, Reading, Mass., 1978.

1.9 Exercises for Chapter 1

1. Suppose $m = 0$ and $n = 0$ and find explicit solutions for $x(t)$, $y(t)$. Consider both the cases when $r, s = 0$ and when $r, s \neq 0$. [Hint : $x'' = ay' = a(bx + s)$.]

2. What happens in arms races in which the null-clines L_1 and L_2 are parallel?

3. Develop a model of arms races involving three countries. How do possible alliances between the three countries impact on the model?

4. Who was Lewis F. Richardson? Write a short synopsis of his life describing how his life experiences influenced his concern about the causes and prevention of wars.

5. Suppose, in Richardson's model, the coefficients a and b are functions of x and y respectively, $a = a(x)$, $b = b(y)$, such that they are positive at $x = 0$ and decrease to 0 at some critical spending level K.

   (a) Find the differential equations arising from this form of a and b. What assumptions would yield such a relationship?

   (b) Repeat the null-cline analysis of your model, either explicitly or computationally.

6. Find a region of the world in which an arms race is occurring or has occurred. Find data about defense spending of the participating power blocks. Estimate the parameters in the model and predict the future of that region. Does Richardson's model really seemed to fit the specific arms race you have chosen?

7. Take, for the parameters a, b, m, n, r, s, the first six numbers of your student ID number. Sketch numerically the phase portrait of the resulting system. Repeat this with the last six numbers. Can you explain any difference you observe in these two phase portraits?

8. Let (x^*, y^*) denote the equilibrium point in Richardson's model. Let

$$u(t) \quad : \quad = x(t) - x^*$$
$$\text{and}$$
$$v(t) \quad : \quad = y(t) - y^*$$

denote the distances to the equilibrium point. Verify carefully that these equations are the same as Richardson's model only with r and s both zero:

$$\begin{aligned} u'(t) &= av(t) - mu(t), \\ v'(t) &= bu(t) - nv(t). \end{aligned}$$

Chapter 2

Phase Portraits of Nonlinear Systems

2.1 Introduction

"There is nothing so practical as a good theory."
  - Kurt Lewin

Making quantitative predictions using models is as simple as calibrating model parameters from data, supplying initial conditions and solving the initial value problem using a reliable numerical method. In spite of the ease and reliability of this procedure, there is still use for exact solutions and good theory. With an exact solution (usually for a simplified problem) one can test how the solution depends on model parameters. With a good theory one can interrogate a model and ask whether its qualitative behavior matches the phenomena modelled. In this chapter we begin to develop such a theory. The theory of this chapter is based on two ideas. The first is representing the global pattern of trajectory directions (the slope field). The second is determining the exact behavior near equilibria by approximation (negotiating with a problem). A problem that cannot be solved is replaced by a nearby one with an exact solution, the linearization. We shall see that in most cases the solution of the nearby, linearized problem is close to the solution of the nonlinear that cannot be solved and describes the local behavior of the nonlinear problem.

We consider the qualitative behavior of nonlinear systems:

$$x' = P(x,y) \ \text{ and } \ y' = Q(x,y). \tag{2.1}$$

We are especially interested in analysis which is local, meaning near an equilibrium point, geometric descriptions which are global, and cases which are mostly of interest in mathematical modeling. Before developing the qualitative behavior, it is important to know that solutions exist. Fortunately, there is a general theorem from ODE theory that asserts existence and uniqueness under mild conditions. We recall next a special case of it.

Theorem 14 *Consider the nonlinear system of ODEs (2.1). Suppose that $P(x,y)$ and $Q(x,y)$ are continuously differentiable. Then for any initial condition*

$$x(0) = x_0, \quad y(0) = y_0$$

there exists a unique solution $x(t), y(t)$ satisfying the initial condition. Further, for any point (x_0, y_0) in the phase plane, there is either exactly one trajectory passing through the point or that point is an equilibrium point.

The assumptions on $P(x,y), Q(x,y)$ for existence and uniqueness to hold can be weakened considerably. That trajectories do not cross in the phase plane uses the fact that the right hand sides $P(x,y), Q(x,y)$ do not depend explicitly on t.

Definition 15 *A system (2.1) is **autonomous** if the right hand sides $P(x,y)$ and $Q(x,y)$ are functions only of x and y, not explicitly of t.*

If a system is autonomous then if $x(t), y(t)$ is a solution then so is $x(t + T), y(t + T)$ for any shift T. To summarize the existence theory:

If the functions P and Q are smooth then, unique solutions exist for any initial condition and, since the system is autonomous, when the solutions are plotted in the phase plane, trajectories cannot cross.

For two equations the phase plane is in the two dimensional $x - y$ plane. In $2d$ non-crossing severely constrains the types of trajectories which can occur in phase portraits of (2.1). In particular, the phase space of (2.1) can only consist of combinations of three things:

- *Critical or equilibrium points.*

- *Non-intersecting trajectories.*

- *Closed curves or cycles which represent periodic solutions of the initial value problem.*

Critical points can be stable or unstable according to the behavior of the trajectories near them.

Definition 16 *If trajectories $x(t), y(t)$ passing near enough to an equilibrium point (x^*, y^*) approach the point as t approaches infinity,*

$$x(t) \to x^* \text{ and } y(t) \to y^*,$$

that equilibrium (x^*, y^*) is called asymptotically stable.

A *cycle* can also be stable or unstable depending on the behavior of trajectories near the cycle.

2.2 Sketching a Phase Portrait by Hand

"Learning to draw is really a matter of learning to see - to see correctly - and that means a good deal more than merely looking with the eye."
 -Kimon Nicolaides

First let us review the procedure we used in the last chapter to sketch phase planes. Suppose we have two ordinary differential equations and we wish to sketch the phase portrait of the system. For example, consider (2.1), an example of which is

$$x' = -2x + y + 4$$

$$y' = -2y + x + 1$$

(2.2)

To sketch their phase portrait, it is important that the functions on the right hand sides of systems (2.1) or (2.2) do not depend explicitly on t, i.e., that the system is autonomous.

The first step is to plot the nullclines of the system, that is, the curves $C_1 : P(x,y) = 0$ where $x'(t) = 0$ and $C_2 : Q(x,y) = 0$ where $y'(t) = 0$. Any trajectory $(x(t), y(t))$ must cross C_1 vertically (since $x'(t) = 0$ on C_1) and C_2 horizontally (since $y'(t) = 0$ on C_2). This is illustrated for (2.2) in Figure 2.1 below.

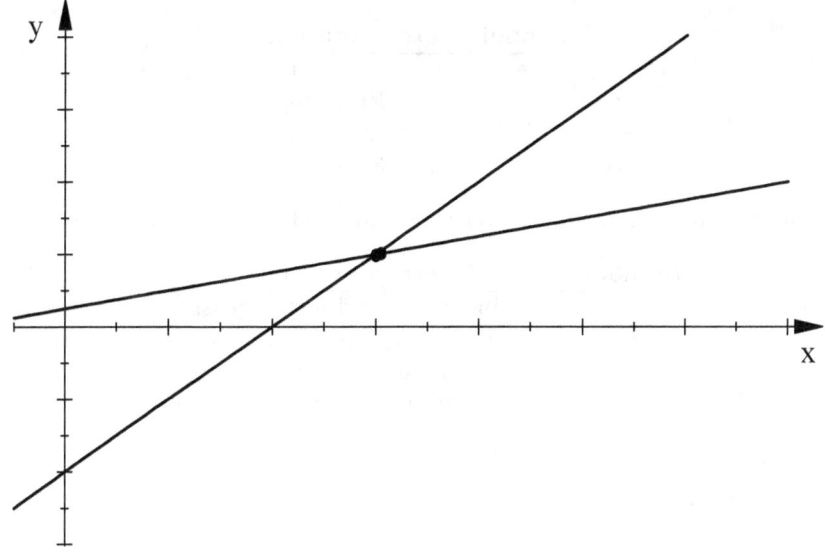

Figure 2.1:   Nullclines C_1 $(-2x + y + 4 = 0)$, C_2 $(-2y + x + 1 = 0)$ for (2.2).
Their intersection is an equilibrium.
Trajectories cross C_1 vertically and C_2 horizontally.

The equilibrium points (x^*, y^*) are the points where $x'(t) = 0$ and $y'(t) = 0$. This is equivalent to

$$P(x^*, y^*) = 0 \qquad and \qquad Q(x^*, y^*) = 0.$$

Specifically, the equilibrium points are where the curves C_1 and C_2 cross. Typically we draw vertical lines on C_1 and horizontal lines on C_2 to remind us that trajectories must cross C_1 vertically and C_2 horizontally.

Now, reconsider the curves C_1 and C_2. C_2 divides the plane into three parts: those points where $y'(t)$ is positive (meaning $y(t)$ is increasing and denoted by an arrow pointing up), those points where $y'(t)$ is negative (meaning $y(t)$ is decreasing and denoted by an arrow pointing down) and those where $y'(t) = 0$ (the curve C_2). Since $y' = Q(x,y)$, the sign of y' on a trajectory passing through a point (x,y) is the sign of Q at the same point (x,y). Evaluating $Q(x,y)$ at one point in each region determines the sign of $Q(x,y)$ on that side of the curve C_2. Next repeat the same analysis for C_1. The curve $C_1 : P(x,y) = 0$ divides the plane into three parts: those points for which $P(x,y)$ is positive (where x' is positive and hence $x(t)$ is increasing, denoted by an arrow pointing to the right), those points for which $P(x,y)$ is negative (where x' is negative and $x(t)$ is decreasing denoted by an arrow to the left) and the curve C_1, where $P(x,y) = 0$. Next, superimpose these figures to get a picture of the trajectory directions $(x(t), y(t))$.

The symbols used to represent trajectory directions have the interpretations described in the table below.

Symbol	Interpretation
$\rightarrow$	x is increasing
$\leftarrow$	x is decreasing
$\uparrow$	y is increasing
$\downarrow$	y is decreasing

Sometimes combinations of the above are represented as below.

Symbol	Interpretation
$\nearrow$	x is increasing and y is increasing
$\searrow$	x is increasing and y is decreasing
$\nwarrow$	x is decreasing and y is increasing
$\swarrow$	x is decreasing and y is decreasing

We now have pictures which includes:

- The nullclines (the curves C_1 and C_2),

- The equilibrium points (where C_1 and C_2 cross),

- The trajectory directions (the arrows in the figures).

This much information is often (but not always, of course) enough to sketch a quantitatively correct phase portrait of (1.1) or (2.1). Remember: trajectories of planar autonomous systems cannot cross! Alternately, you can simply plot the vector field $(P(x,y), Q(x,y))$. For the system (2.2) this is the vector field $(-2x + y + 4, -2y + x + 1)$, plotted next in Figure 2.2. When doing so, choices

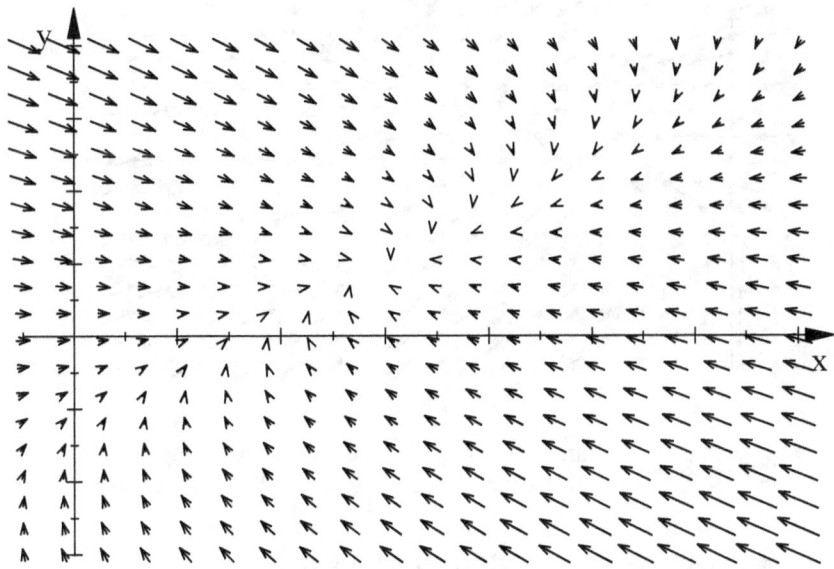

The vector field $(-2x + y + 4, -2y + x + 1)$.

Figure 2.2:   The vectors are not scaled to have the same length.
(Often rescaled vectors are plotted.)

must be made that affect how clear the vector field represents trajectory directions. The two main ones is how dense should the arrows be and whether the vectors length is fixed or represents the magnitude of the vector. The last plot scales the arrows by vector length. The next plot, Figure 2.3, is the same vector field where the plotting program is given a scaled vector field to plot. Here the program is given a vector field with all vectors having length $= 1$. (Obviously, the program erases some parts of vectors that would lie over another vector.)

For example, for the system (2.1) we can draw in a few representative trajectories as sketched[1] below in Figure 2.4. The plot shows that the equilibrium point of (2.1) is stable: all trajectories ultimately approach the equilibrium as t approaches infinity.

---

[1] This plot is not hand drawn. Sadly, hand drawn plots are frowned upon in publishing even though they can be more useful for learning.

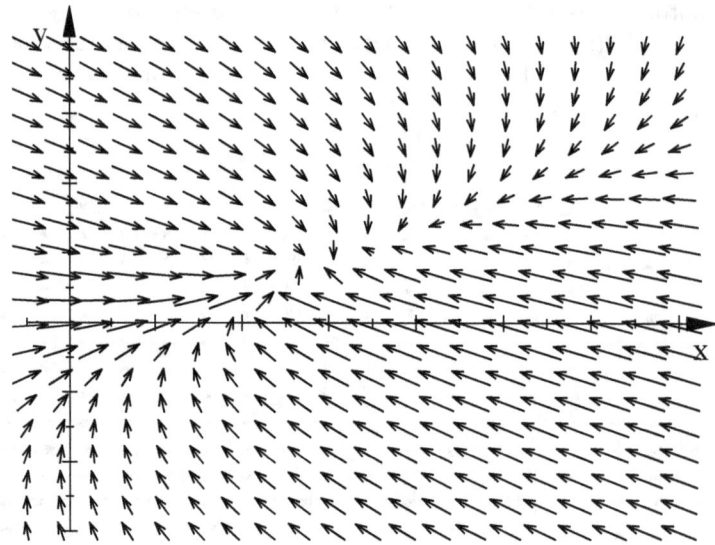

Figure 2.3:   Same vector field: rescaling vector length makes the directions near an equilibrium clearer.

2.3 Analysis Near Critical Points

The previous section looked at graphical representation of the global phase portrait of a nonlinear system:

$$
\begin{aligned}
x' &= P(x,y), \\
y' &= Q(x,y).
\end{aligned}
\tag{2.3}
$$

In this section we look at the local behavior of trajectories near a critical point (x^*, y^*). The analysis near a critical point is based on linearization of the system. The functions $P(x,y)$ and $Q(x,y)$, near (x^*, y^*), are arbitrarily close to their linearization. Since the linearized problem can be solved exactly we can sketch the phase portrait of the linear problem.

First by shifting we may assume the equilibrium point is at $(0,0)$. Indeed, writing

$$u = x - x^* \text{ and } v = y - y^*$$

we obtain

$$
\begin{aligned}
Q^*(u,v) &= Q(u + x^*, v + y^*) \\
P^*(u,v) &= P(u + x^*, v + y^*).
\end{aligned}
$$

Thus, we find that the shifted functions $u(t)$ and $v(t)$ satisfy

$$
\begin{aligned}
u' &= P^*(u,v) \\
v' &= Q^*(u,v),
\end{aligned}
$$

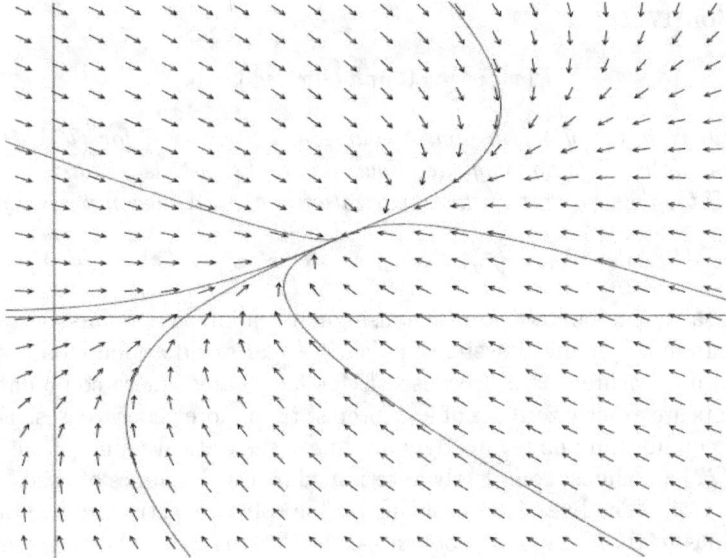

Figure 2.4:   The phase portrait of (2.2).
The equilibrium is called a stable node.

where

$$P^*(0,0) = 0 \text{ and } Q^*(0,0) = 0,$$
$$\text{and}$$
$$P_x^*(0,0) = P_x(x^*,y^*) \text{ and } P_y^*(0,0) = P_y(x^*,y^*),$$
$$Q_x^*(0,0) = Q_x(x^*,y^*) \text{ and } Q_y^*(0,0) = Q_y(x^*,y^*).$$

Thus, expanding P^* and Q^* in a Taylor series about $(0,0)$ gives the same linearized problem as expanding P, Q about (x^*,y^*). (This means we can drop the notation P^* and Q^*.) The system is

$$\begin{cases} u' &= P_x(x^*,y^*)u + P_y(x^*,y^*)v, \\ v' &= Q_x(x^*,y^*)u + Q_y(x^*,y^*)v. \end{cases} \quad (2.4)$$

We shall denotes by A the coefficient matrix on the right hand side of (2.4):

$$A = \left[\begin{array}{cc} P_x(x^*,y^*) & P_y(x^*,y^*) \\ Q_x(x^*,y^*) & Q_y(x^*,y^*) \end{array} \right]$$

The equations (2.4) are linear, constant coefficient, homogeneous ordinary differential equations, which can therefore be solved exactly. We shall restrict our attention to nonsingular, sometimes called "isolated" critical points, defined as follows.

Definition 17 *Let*

$$P(x^*, y^*) = 0 \ and \ Q(x^*, y^*) = 0.$$

The equilibrium (x^*, y^*) is a nonsingular equilibrium point for (??) if the linearization matrix A is nonsingular. Equivalently, (x^*, y^*) is a nonsingular equilibrium if the determinant of the linearization matrix A does not vanish:

$$det(A) := P_x(x^*, y^*)Q_y(x^*, y^*) - P_y(x^*, y^*)Q_x(x^*, y^*) \neq 0.$$

We shall focus on the case of nonsingular equilibrium points because the coefficients in a dynamical system modeling a real world application are determined from measurements. Because of this fact, cases where combinations of coefficients are exactly zero are of less interest than more generic cases. For nonsingular equilibrium points, the dynamics near the critical point (x^*, y^*) of the system (??) are almost completely determined by the dynamics of the linearized problem (2.4). The linearized problem can be solved exactly and the behavior of solutions to the linearized problem can be determined by the eigenvalues of the linearization matrix A.

Let us now present the relevant cases. These phase portraits are sketched by diagonalizing the matrix A and then solving explicitly the 2×2 linear constant coefficient, homogeneous, ordinary differential equation.

Case 18 (Node)

$$A = \begin{bmatrix} a & 0 \\ 0 & b \end{bmatrix},$$

where $\lambda = a, b$ are the eigenvalues of matrix A, with the same sign.

For $0 < a < b$ the slope field near $(0, 0)$ and a representative phase plane near $(0,0)$ are plotted below in Figure 2.5 (the slope field) and Figure 2.6 (the phase portrait). When the signs are reversed, $b < a < 0$, the picture is similar but with reversed arrows (and trajectory directions). When $0 < a < b$ the equilibrium is unstable and when $b < a < 0$ it is stable. When the magnitudes of a and b are reversed the picture is rotated 90 degrees. If $a = b$ then the point is called a (stable or unstable) star point. Star points are not the generic case. When the model's coefficients come from data and are thus known only to finite precision, exact equality is a fluke. An example of a star point is plotted next in Figure 2.7 for $a = b > 0$.

Case 19 (Saddle)

$$A = \begin{bmatrix} a & 0 \\ 0 & b \end{bmatrix},$$

where $b < 0 < a$. In a saddle point, if the initial condition is an exact multiple of the negative eigenvector, then the solution approaches the equilibrium as $t \to \infty$. This line (all multiples of the eigenvector of the negative eigenvalue) is called the stable manifold. Similarly, the unstable manifold is the line that consists of all multiples of the eigenvector of the positive eigenvalue. For the diagonalized

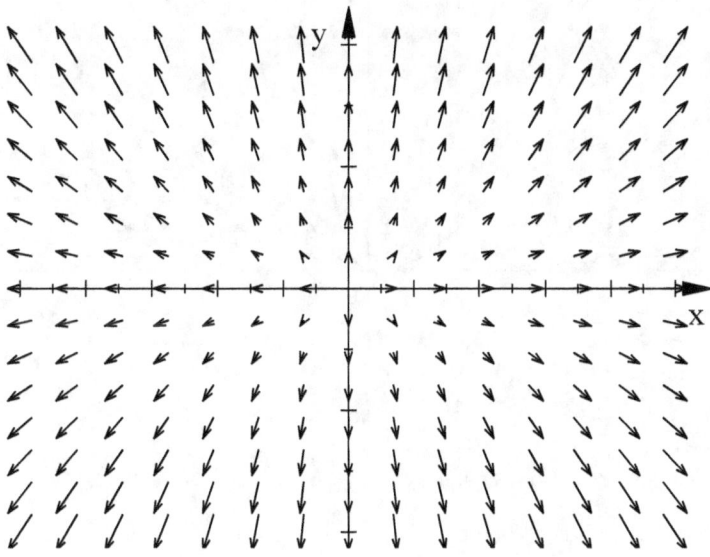

Figure 2.5: Slope field near (0,0), an unstable node here.

matrix (above) these are the x and y axes. For an example we plot the phase portrait of $x' = x, y' = -y$ eigenvalues +1 and −1) next in Figure 2.8. For more general, nonlinear problems, the stable and unstable manifolds are curves that are tangent at the equilibrium to the lines that are the linearizations stable and unstable manifolds. The lines do not have to be orthogonal as depicted above.

Case 20 (Spiral) *Complex eigenvalues $a+ib$ and $a-ib$. This is a stable spiral if $a < 0$ and an unstable spiral if $a > 0$.*

Case 21 $a = b$ *(both real) and there is only one eigenvector.*
This case is not generic: any small perturbations of model data would shift the matrix and its eigenvalues so that $a \neq b$. Thus it is not of primary interest for models of phenomena. Here

$$A = \begin{bmatrix} a & 1 \\ 0 & a \end{bmatrix},$$

so the solution to the linearized system:

$$\begin{aligned} u' &= au + v \\ v' &= av, \end{aligned}$$

is

$$\begin{aligned} u(t) &= (v(0)t + c)e^{at} \\ v(t) &= v(0)e^{at}. \end{aligned}$$

This case is called an improper node, depicted in Figure 2.10 below. It is stable if $a < 0$ and unstable if $a > 0$.

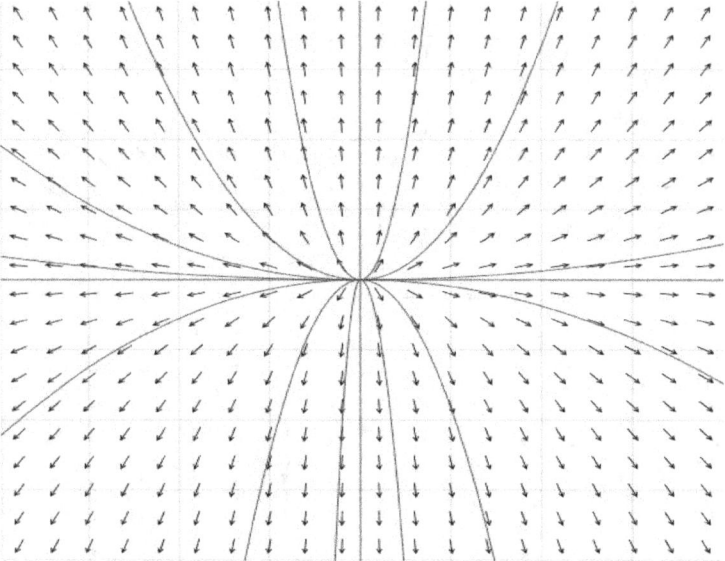

Figure 2.6: Phase portrait when $0 < a < b$, an unstable node

Case 22 *Purely imaginary eigenvalues $\lambda = +ai$ and $\lambda = -ai$.*
This case is also not generic but is nevertheless important as a transition between two types of models that are of importance. In this case the linearized problem has solutions

$$
\begin{aligned}
u(t) &= C\cos(at + w) \\
v(t) &= C\sin(at + w),
\end{aligned}
$$

for some constants C and w. These solutions trace out circles: $u^2 + v^2 = C^2$. This case is called a linear center; we shall see that the nonlinear problem may be a center or a spiral. A classic example of a linear center is the system

$$
x' = +y \ \& \ y' = -x.
$$

For this system the matrix is

$$
\begin{bmatrix} 0 & +1 \\ -1 & 0 \end{bmatrix} \text{ with eigenvalues } \lambda_1 = +i, \lambda_2 = -i.
$$

A phase portrait of a linear center is presented next in Figure 2.11.

2.4   The Nonlinear Problem

Recall that the phase space of

$$
\begin{cases} x' = P(x,y), \\ y' = Q(x,y). \end{cases} \tag{2.5}
$$

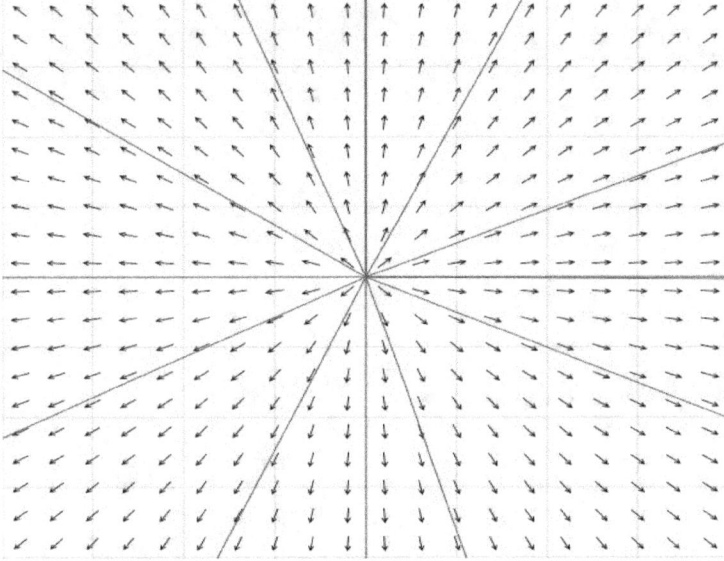

Figure 2.7: (Unstable) Star Point $a = b(> 0)$

must be composed of combinations of only three things:

- *critical points or equilibria.*

- *Non-intersecting trajectories.*

- *Closed curves or cycles.*

The behavior of the nonlinear problem near a critical point can be *nearly* completely analyzed by solving the linearized problem exactly. It is remarkable that the behavior of this linearized problem almost (all except case f below) completely determines the behavior of the nonlinear problem there.

Theorem 23 *Suppose $P(x,y)$ and $Q(x,y)$ are smooth enough (In particular, that their second derivatives are continuous). Then the following hold.*

(a) *If an equilibrium point is asymptotically stable or unstable for the linearized problem then it is also for the nonlinear problem as well.*

(b) *If the equilibrium point is a spiral point for the linearized problem then it is also a spiral point for the nonlinear problem.*

(c) *If the equilibrium point is a node for the linearized problem then it is a node for the nonlinear problem also.*

(d) *If the equilibrium point is a saddle point for the linearized problem then it is a saddle point for the nonlinear problem also.*

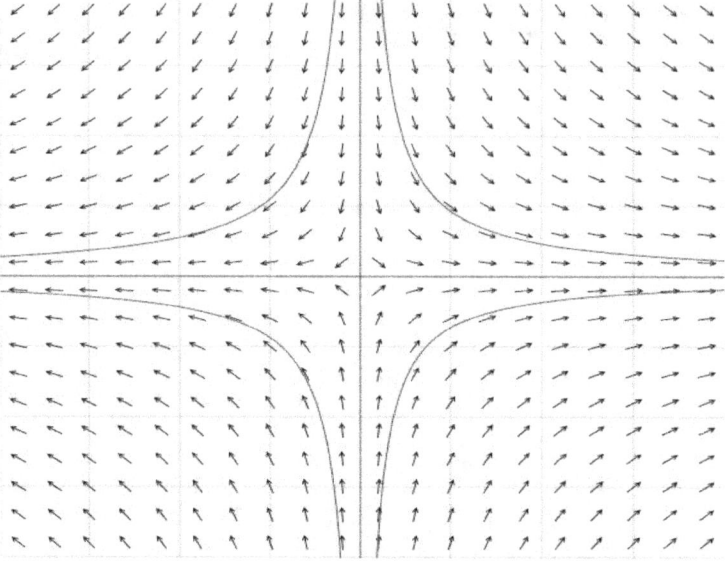

Figure 2.8:   An example of a saddle point: $x' = x, y' = -y$.
Here the x-axis is the ustable manifold & y-axis the stable manifold.

(e) *If the equilibrium point is a proper node for the linearized problem then it
is also a proper node for the nonlinear problem.*

(f) *If the equilibrium point is a center for the linearized problem then it is
either a center, a stable spiral, or an unstable spiral for the nonlinear
problem.*

The previous theorem insures that the local behavior near critical points
/ equilibria is usually determined (and determinable from) by that of the linearization. However, it must be understood that this theorem speaks only to
local behavior. Away from critical points the nonlinear terms in the system
strongly influence the trajectories. Consider the example (2.6) that follows.

$$\begin{aligned} x' &= x + 4y + e^x - 1, \\ y' &= -y - ye^x. \end{aligned} \qquad (2.6)$$

This system has a critical point at $(0,0)$. The linearized system at $(0,0)$ is easily
calculated to be

$$\begin{aligned} u' &= 2u + 4v, \\ v' &= -2v. \end{aligned} \qquad (2.7)$$

The linearization matrix A is thus given by

$$\begin{bmatrix} 2 & 4 \\ 0 & -2 \end{bmatrix}.$$

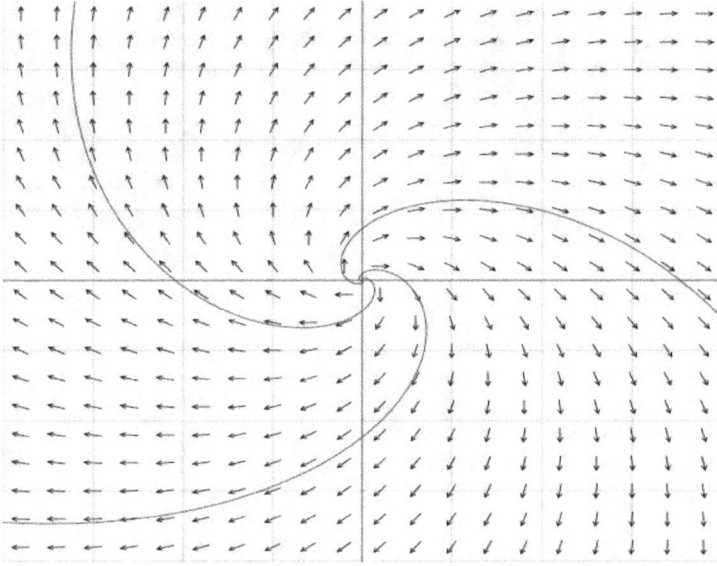

Figure 2.9: (Unstable) Spiral

Its eigenvalues are easily calculated to be $\lambda = \pm 2$. Indeed, for this matrix A

$$\det\{A - \lambda I\} = \det \begin{bmatrix} 2 - \lambda & 4 \\ 0 & -2 - \lambda \end{bmatrix} = (2 - \lambda)(-2 - \lambda) = 0$$

which implies

$$\lambda = \pm 2.$$

This example is a special case of the following general result from linear algebra.

Proposition. *If A is an $N \times N$ upper triangular ($a_{ij} = 0$ if $i > j$) matrix then the eigenvalues of A lie in the diagonal:*

$$\lambda_i = a_{ii}, i = 1, \cdots, N.$$

proof: *The proof is quite similar to the above example. For the $N \times N$ case the only added feature is to evaluate $\det\{A - \lambda I\}$ by expansion by cofactors down the first column. The $N - 1 \times N - 1$ determinant that then arises is expanded down its first column. With this hint, we leave the rest of the formal proof as an exercise.*

The associated eigenvectors of A are also easy to calculate and determine the stable and unstable manifolds[2] of the linearization. The stable and unstable manifolds of the nonlinear problem are tangent to these at the equilibrium. The eigenvectors and associated manifolds are:

---

[2] "Manifold" has manifold meanings. As used here, it is just a curve.

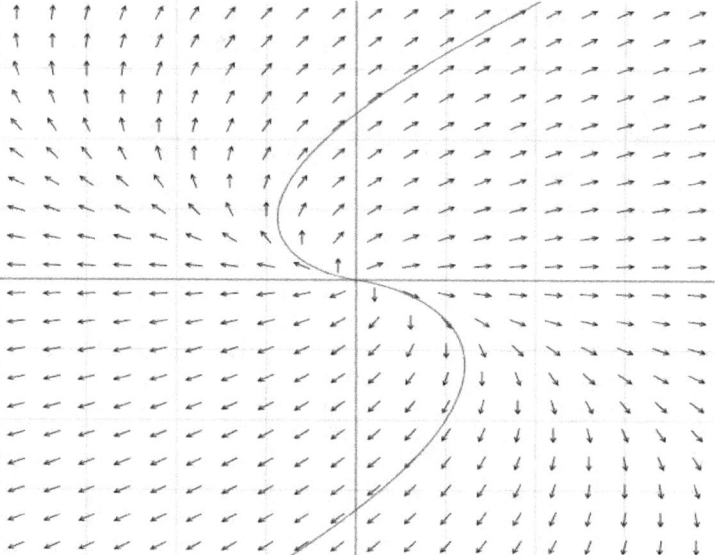

Figure 2.10: An example of an improper node

λ	eigenvector	manifold = eigenspace	
2	$\begin{bmatrix} 1 \\ 0 \end{bmatrix}$	$\begin{bmatrix} t \\ 0 \end{bmatrix}$	(unstable)
-2	$\begin{bmatrix} -1 \\ 1 \end{bmatrix}$	$\begin{bmatrix} -t \\ t \end{bmatrix}$	(stable)

Thus, $(0,0)$ is a saddle point for both the linearized problem (2.7) and the nonlinear system (2.6). In the next two figures, Figure 2.12 and 2.13, we give the phase portraits for both the nonlinear problem and its linearization.

The stable manifold (the trajectory that approaches the equilibrium as $t \to \infty$) and the unstable manifold (the trajectory approaching the equilibrium as $t \to -\infty$) of the linearized and nonlinear problem are tangent to one another at the critical point, compare the above phase portraits of the linearized and nonlinear problem. The farther one goes from the point of linearization, the greater the difference between the dynamics of the linear and nonlinear problems. Global dynamics are not determined by behavior near equilibrium points. This is universal; in the small, linear terms dominate nonlinear ones and in the large nonlinear terms dominate linear terms. In its simplest form:

$$\text{for (small) } |x| < 1, \quad |x^2| < |x|, \quad \text{e.g., } \left(\tfrac{1}{100}\right)^2 << \tfrac{1}{100},$$
$$\text{for (large) } |x| > 1, \quad |x^2| > |x|. \quad \text{but } 100^2 >> 100.$$

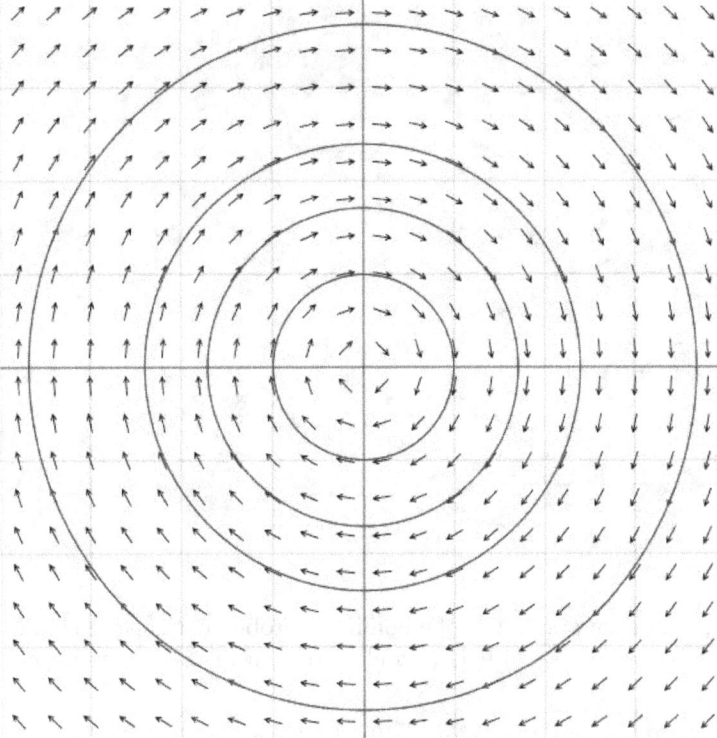

Figure 2.11: An example of a (linear) center.

2.5 Purely analytical approaches fail

"... many branches of both pure and applied mathematics are in great need of computing instruments to break the present stalemate created by the failure of the purely analytical approach to nonlinear problems." - J. von Neumann, 1945.

Exact paper and pencil analysis is limited to linear problems and a few odd examples where closed form solutions are possible using a small number of special tricks. Linear problems, as we have seen, give insight into behavior near critical points. Exact solutions are used, correctly or incorrectly, to suggest trends in more general nonlinear problems. This situation was the basis of the above assessment of von Neumann. The advance in understanding of nonlinear problems since 1945 is based on the vision of von Neumann.

Dynamical systems (herein) do not need the full power of computers: the planet Neptune was discovered by J.C. Adams using numerical methods where all the calculations were done by hand. Nevertheless, advanced numerical methods, especially self-adaptive methods, have revolutionized the modern world and are very useful even for ODE models. We use two convenient numerical tools

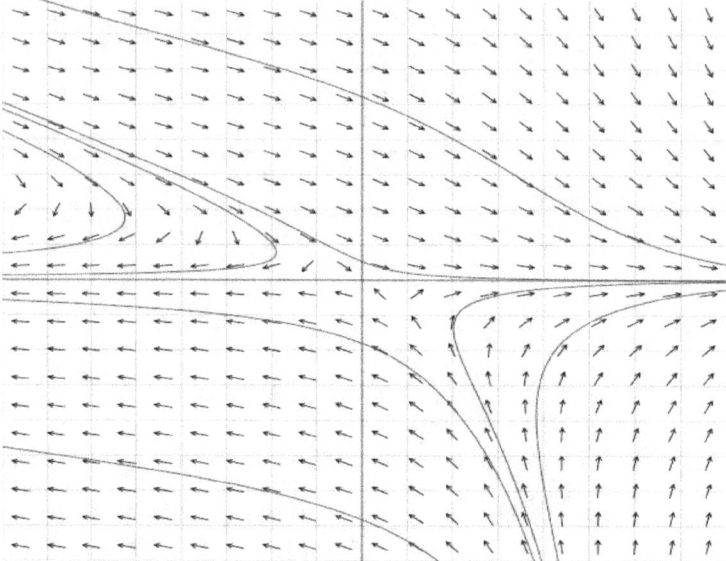

Figure 2.12:   Phase portrait of the nonlinear problem: Near the equilibrium at (0,0) It is very close to that of the linearization.

herein.

Visualization of slope fields.

"All sorts of computer errors are now turning up. You'd be surprised to know the number of doctors who claim they are treating pregnant men." - Anonymous Official   of   the   Quebec   Health   Insurance   Board

The slope field or vector field of an autonomous system of two equations

$$x' = P(x,y), \ y' = Q(x,y)$$

is a vector $(P(x,y), Q(x,y))$ at each point (x,y) in the plane. This vector field is easily visualized by choosing a discrete grid of points and plotting the vector at each point. As an example, for

$$\begin{aligned} x' &= x + 4y + e^x - 1, \\ y' &= -y - ye^x, \end{aligned} \tag{2.8}$$

the slope field is $(x + 4y + e^x - 1, -y - ye^x)$ plotted below on a 10×10 grid. In the left and right figures the arrow length of the vector field given to the plotting program is, respectively,

Left Figure: arrow length $= \sqrt{(x + 4y + e^x - 1)^2 + (-y - ye^x)^2}$,
Right Figure: arrow length $=$ 1.

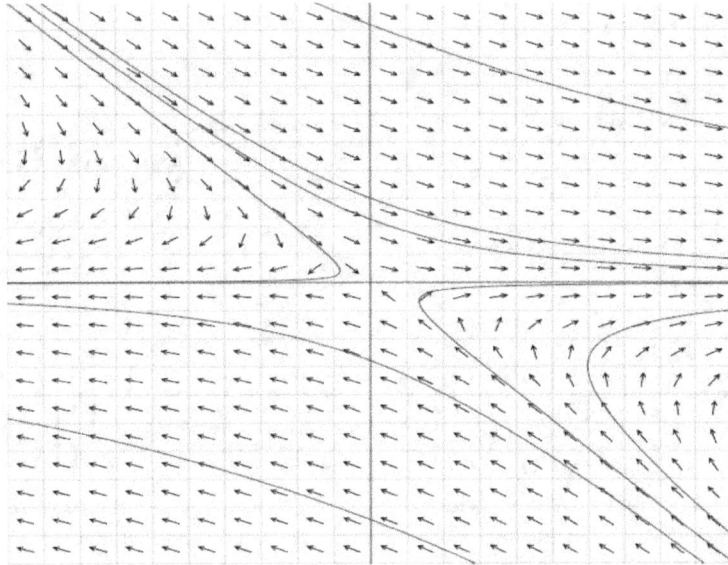

Figure 2.13: Phase portrait of the linearization. Near (0,0) It is very close to the nonlinear problem's portrait.

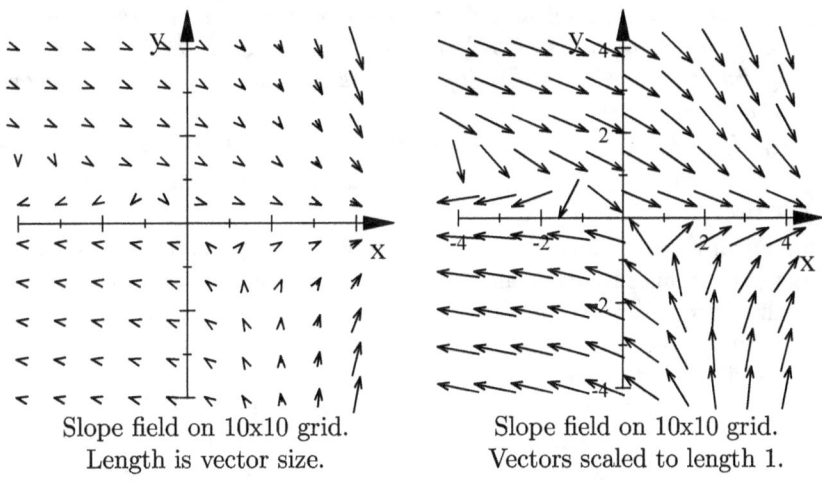

Slope field on 10x10 grid. Length is vector size.

Slope field on 10x10 grid. Vectors scaled to length 1.

With the plotted slope field, one can make a reasonable guess about trajectories simply by drawing curves tangent to the vectors and in their direction. The critical question is *How do we know this is correct?* There could well be complicated behavior on smaller scales than represented on the grid. One can repeat on a finer grid, as below, pushing the same question to smaller scales.

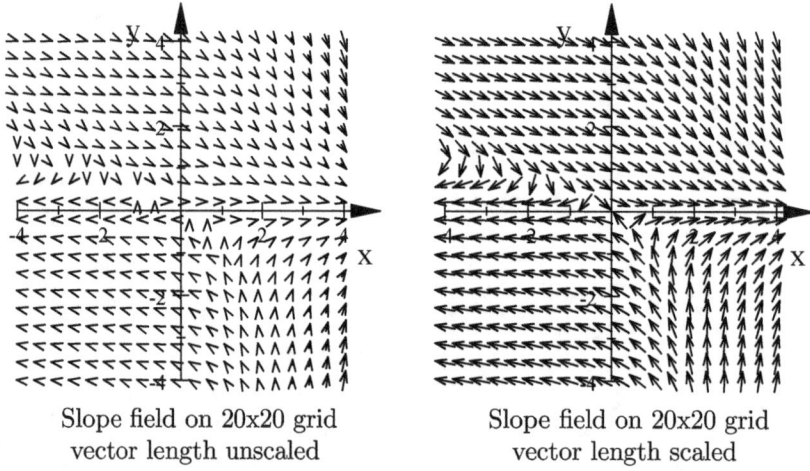

Slope field on 20x20 grid   Slope field on 20x20 grid
vector length unscaled   vector length scaled

One common rule of thumb for nonlinear problems is that if the same picture appears for 3 successive doublings of the resolution, then its accuracy is plausible, until checked by a self-adaptive numerical method.

Numerical Methods for Initial Value Problems

"Perhaps the history of the errors of mankind, all things considered, is more valuable and interesting than that of their discoveries. Truth is uniform and narrow; it constantly exists, and does not seem to require so much an active energy as a passive aptitude of the soul in order to encounter it. But error is endlessly diversified; it has no reality, but it is the pure and simple creation of the mind that invents it."

- Benjamin Franklin, Report of Dr. B. Franklin and other commissioners, Charged by the King of France with the examination of Animal Magnetism, as now practiced in Paris, 1784.

Numerical methods for solving initial value problems were used by Adams to discover the planet Neptune and by NASA to predict the future trajectory of space vehicles to control their travel to the moon and back. **Reliability** means the user inputs the problem data, the parts of the solution needed and the number of significant digits needed (the error tolerance) and the algorithm self-adjusts to provide what is needed to the accuracy required. **Efficiency** means that this prediction is made with minimal or near minimal cost / turnaround time / resources. These are competing demands. Assured accuracy means a small time step is desirable while for fast turn around the total number of timesteps should be minimized.

Suppose the state of a system at time t can be characterized by a collection of N numbers (the vector $y(t)$). N is often quite large but $N = 2$ for most of the models herein. Suppose the state of the system is known today (taken to be $t = 0$) and finally that the laws governing the system are known: the way the system changes depends on the time and on the state of the system

$$y' = f(t, y), \text{ for all time } t > 0 \qquad \text{(IVP)}$$
$$y(0) = y_0, \text{ at time } t = 0.$$

The *initial value problem* is then to predict (reliably and efficiently) the future state of the system: find $y(t)$ for $t > 0$.

The most basic solution to this problem was devised by the great Leonard Euler. It (**Euler's method**[3]) proceeds as follows: pick a step size called $\triangle t$. The variables t_j and y_j denote $t_j = j \triangle t$ and y_j is the approximation we compute to $y(t_j)$:

$$\triangle t = \text{step size}, \quad t_j = j \triangle t = j^{th} \text{ time step}, \quad y_j \approx y(t_j).$$

Euler's method to find y_j is constructive. It is motivated as follows: Suppose we know $y(t_j)$ exactly and want $y(t_{j+1}) = y(t_j + \triangle t)$. Expanding y in a Taylor series at t_j gives:

$$y(t_{j+1}) = y(t_j) + y'(t_j)\triangle t + \frac{1}{2}y''(\xi)\triangle t^2 \text{ , for some } \xi, \ t_j < \xi < t_{j+1}.$$

Now the equation $y(t)$ satisfies is $y'(t_j) = f(t_j, y(t_j))$. Thus:

$$y(t_{j+1}) = y(t_j) + \triangle t f(t_j, y(t_j)) + \frac{1}{2}y''(\xi)\triangle t^2 \text{ , for some } \xi, \ t_j < \xi < t_{j+1}.$$

The last term, $\frac{1}{2}y''(\xi)\triangle t^2$, is unknowable since both y'' and the point ξ are unknown but the term is small[4] if $\triangle t$ is small. Just dropping this last term is Euler's method:

$$\text{Given } y_j,$$
$$\text{Find } y_{j+1} \text{ by}$$
$$y_{j+1} = y_j + \triangle t f(t_j, y_j), \qquad \text{(Euler)}$$
$$\text{for } j = 0, 1, 2, \cdots.$$

Rearranging $y_{j+1} = y_j + \triangle t f(t_j, y_j)$, this is equivalent to

$$\frac{y_{j+1} - y_j}{\triangle t} = f(t_j, y_j), \ y_j \text{ given}.$$

In theory this works: It is easy to program and constructive. It is known that the method also converges to the exact solution

$$y_n \to y(t_n) \text{ as } \triangle t \to 0,$$

---

[3] Using Euler's method rather than trying to find an exact, closed form solution to a system of ODEs was the key breakthrough in the movie *Hidden Figures*.

[4] In applied math, something is small usually means that something^2 is negligable.

so that in theory by taking $\triangle t$ small enough whatever is needed can be obtained. In practice however, Euler's method is nearly completely inadequate. It has low accuracy: on many computers the small amount of roundoff errors always present accumulates fast enough to overwhelm the method's accuracy. Its error can grow exponentially fast as more steps are taken even in cases when the true solution does not. The approximate solution can also grow rapidly in cases when the true solution does not. This is known as an instability and it can happen many different ways. For one example, consider

$$y' = -10y, \text{ and } y(0) = 1.$$

The true solution is $y(t) = e^{-10t}$ which simply goes to zero rapidly.

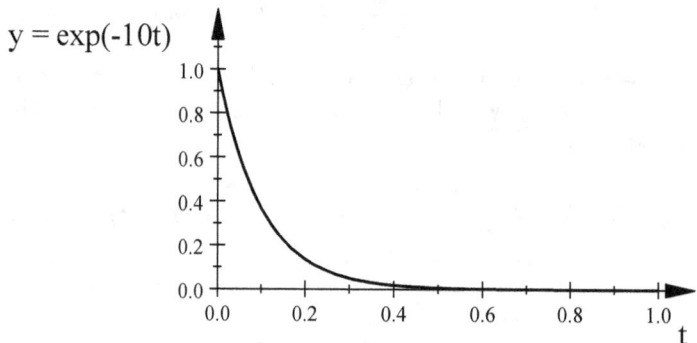

The solution $y(t) = e^{-10t} \to 0$ very fast.

Euler's method for this (simple) test problem has an exact solution

$$\frac{y_{j+1} - y_j}{\triangle t} = -10y_j \text{ so } y_j = (1 - 10\triangle t)^j .$$

If, for example, $\triangle t = 0.3$ then $y_j = (-2)^j$, a clear instability, Figure 2.14.

Fortunately, there have been great advances since the time of Euler in both methods of greater accuracy and better stability and in adaptive error control. These modern methods give reliable predictions and can generate phase portraits reliably. For example, for the system analyzed in the last section

$$\begin{aligned} x' &= x + 4y + e^x - 1, \\ y' &= -y - ye^x, \end{aligned} \tag{2.9}$$

the phase portrait in Figure 2.15 below was generated completely by a numerical method. The trajectories were plotted by solving the initial value problem with different initial conditions and adaptive timesteps.

2.6   References for Chapter 2

There are many routines for plotting phase portraits numerically. Almost all are free and easy to use. There are even many online where you type in the

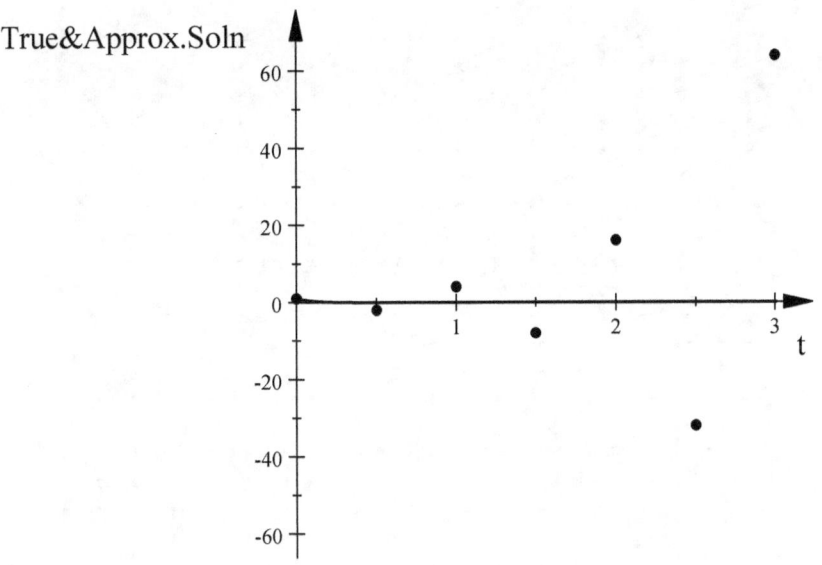

Figure 2.14:
$$y' = -10y : \text{ True } y(t_j) \to 0 \text{ but}$$
Euler approx.: $y^j = (-2)^j$ blows up & oscillatees.

equation and the plot is made. The currently available ones are easily found by a web search and change from year to year. Other options include:

1. Professor J.C. Polking's routines at: http://math.rice.edu/~dfield/dfpp.html are good and widely used.

2. Mathematica (The equation trekker feature), Maple and MATLAB can all do it easily.

4. Here is an example of MATLAB code for the system $x' = x(7-x-2y), y' = y(5 - 3x - 4y)$:

```
% domain is [-5,5] x [-5,5] 50x50 subintervals
xdom = linspace(-5,5,51);
ydom = linspace(-5,5,51);
[X,Y] = meshgrid(xdom,ydom);
U = X*(7.0 - X - 2.0*Y); % dx/dt
V = Y*(5.0 - 3.0*X - 4.0Y); % dy/dt
quiver(X,Y,U,V)
```

3. If you search the web on the term 'phase plane plotting', a number of sites will appear that have routines that can be used through a web browser interface. These are constantly changing so look up a few and experiment! The plots herein were made variously using MATLAB, applets, Polking's program, and muPAD[5] (a computer algebra program). The number available on the web is constantly changing.

---

[5] This book was composed using Scientific Workplace which includes muPAD.

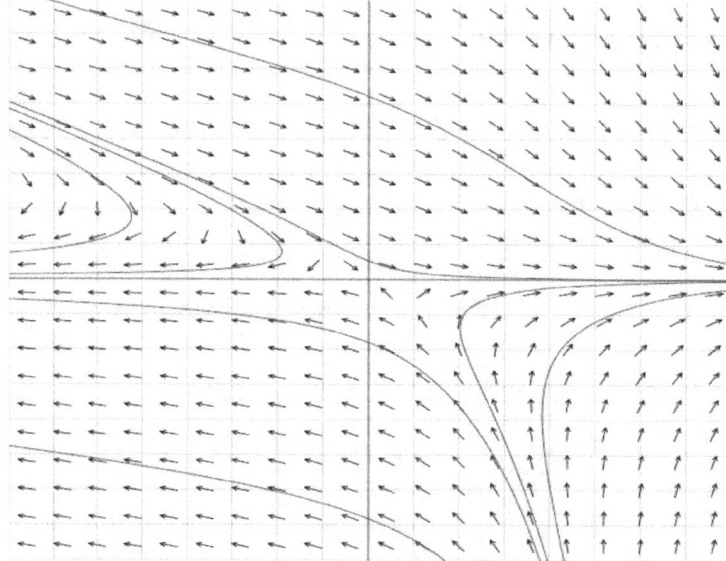

Figure 2.15: Phase portrait: $x' = x + 4y + e^x - 1$, $y' = -y - ye^x$

There are also several books which come with an easy to use program for plotting phase portraits.

D. Arnold and J.C. Polking, Ordinary Differential Equations using MAT-LAB, Prentice Hall, 1999.

J. H. Hubbard and B. H. West, MacMath 9.0, A dynamical Systems Package for the MacIntosh, Springer, Berlin, 1992.

H. Kocak, Differential and Difference Equations through Computer Experiments, Springer-Verlag, Berlin, 1989.

2.7   Exercises for Chapter 2

1. Consider the system

$$\begin{aligned} x' &= y, \\ y' &= 2x - x^2 \end{aligned}$$

Sketch the nullclines, trajectory directions and equilibria. Sketch a few representative trajectories. Compare your sketch with a computer generated phase portrait if possible.

2. Repeat problem 1 for the system

$$\begin{aligned} x' &= -4x + 2xy - 8, \\ y' &= 4y^2 - x^2 \end{aligned}$$

3. Repeat problem 1 for the system

$$\begin{aligned} x' &= x^2 - y^2 - 1, \\ y' &= 2y \end{aligned}$$

4. Sketch the phase portraits near $(0,0)$ of the following systems either analytically or computationally.

a. $x' = -x, y' = -y$.

b. $x' = -x, y' = -2y$.

c. $x' = x, y' = -1$.

d. $x' = y, y' = -x$.

e. $x' = -x, y' = -x + y$.

5. For some systems of two ODEs one can eliminate one variable and reduce two first order equations to one second order equation. For example: in (d) $x' = y$ so $x'' = y' = -x$ satisfies a second order equation $x'' = -x$. Which of the examples from the last exercise might have come from a second order scalar equation?

6. Find and classify the critical points of the following systems. Sketch their phase portraits (with computer assistance if desired).

(a)
$$\begin{cases} x' &= 3x + 4y, \\ y' &= -3x - 3y. \end{cases}$$

(b)
$$\begin{cases} x' &= x, \\ y' &= y^2. \end{cases}$$

(c)
$$\begin{cases} x' &= x^2, \\ y' &= y(2x - y). \end{cases}$$

7. Consider the nonlinear system

$$\begin{cases} x' &= \sin x, \\ y' &= -\sin y. \end{cases}$$

Find and classify the critical points under the condition $-2\pi < x, y < 2\pi$. Sketch the system's phase portrait.

8. Consider the equilibria of the nonlinear pendulum equation with a parameter a:

$$\theta'' + a\theta' + \sin \theta = 0.$$

(a) Write it as a first order system and sketch the phase portraits for a value of the parameter $a < 0, a = 0$ and $a > 0$.

(b) You will observe that when $a = 0$ the equilibrium $(0,0)$ is a center. By choosing small but nonzero values of the parameter a, investigate experimentally the stability of the center at $(0,0)$ in the first equation to small changes.

(c) Try to formalize a conclusion from your experiments.

9. Consider the critical point $(0,0)$ of the system

$$\begin{cases} x' & = & -y - x\sqrt{x^2 + y^2}, \\ y' & = & x - y\sqrt{x^2 + y^2}. \end{cases}$$

(a) Find the linearized problem and show the origin is a center for the linearized problem.

(b) Convert this system to a system in polar coordinates for $\theta(t)$ and $r(t)$.

(c) Show $r(t) \to 0$ as $t \to \infty$ thereby showing that the origin is in fact a spiral point for the nonlinear system.

Hint: As $r^2 = x^2 + y^2$, differentiating with respect to t via the chain rule gives

$$2rr' = 2xx' + 2yy'.$$

Insert the equations for x' and y' and simplify. Similarly, $\tan\theta = y/x$. Differentiate with respect to t, substitute for x' and y' and simplify.

10. Find the eigenvalues and eigenvectors of the 2x2 matrix below. Using that information [and possibly consulting a book on ODEs] solve the associated system:

$$A = \begin{bmatrix} -3 & 4 \\ 4 & 3 \end{bmatrix}$$

System:

$$x' = -3x + 4y,$$
$$y' = 4x + 3y.$$

11. Consider

$$x' = (2 - x - 2y)x,$$

$$y' = (2 - 2x - y)y.$$

Find all 4 critical points. Consider the critical point $(2/3, 2/3)$. Find the linearization there and its eigenvalues. Sketch the phase portrait *of the linearization* in the variables

$$u = x - 2/3$$
$$\text{and}$$
$$v = y - 2/3.$$

12. [A computational exercise] The theory says that linear centers need not be stable to higher order but small nonlinear perturbations. Test this numerically and make a catalog of examples of what can happen to a linear center under a small perturbation. You are free and encouraged to construct your own examples. To give you an idea how it can be done, one way is to start with $x' = y, y' = -x$. Then add terms like

$$\varepsilon(ax^2 + bxy + cy^2)$$

where a, b, c are $O(1)$ and ε is small and then see what happens for different choices of a, b, c, ε.

13. Repeat problem 1 for the system

$$x' = x - y^3,$$

$$y' = y + x^3.$$

Additionally, classify the critical points.

14. Repeat problem 1 and classify the critical points. for the system

$$x' = 1 - x^2 - y^2,$$
$$y' = y - x$$

15. Find and classify all critical points of

$$(a)\ x' = -4y + 2xy - 8,$$
$$y' = 4y^2 - x^2$$
$$(b)\ x' = x^2 - y^2 - 1,$$
$$y' = 2y$$
$$(c)\ x' = 1 - x^2 - y^2,$$
$$y' = y - x.$$

16. [Fill in the blanks] Consider the system

$$(*)\ x' = P(x, y)\ \text{and}\ y' = Q(x, y).$$

Then, the phase space of $(*)$ consists only of _____.
The equilibrium points of $(*)$ are _____.
An equilibrium points of $(*)$ is nonsingular if _____.
The nullclines of $(*)$ are _____.

17. Consider the system

$$x' = P(x, y),\quad y' = Q(x, y)$$

which has a simple critical point at (x^*, y^*). This problem is about changing variables to polar coordinates around the critical point by

$$\begin{aligned} x &= x^* + r\cos(\theta) \\ y &= y^* + r\sin(\theta). \end{aligned}$$

(a) [This part uses the formula for the inverse of a 2×2 matrix.] Show that the $r(t)$ equation in the system becomes

$$r'(t) = F(r, \theta) := \cos(\theta)P(r\cos(\theta), r\sin(\theta)) + \sin(\theta)Q(r\cos(\theta), r\sin(\theta)).$$

(b) [This part is an epsilon - delta proof] Prove: If there is an $\varepsilon_0 > 0$ such that for all $0 < r < \varepsilon_0, 0 < \theta \le 2\pi$ there holds $F(r, \theta) < 0$ then the critical point is stable.

18. Consider the system of ODEs:

$$\begin{aligned} x' &= 1 - x^2 - y^2, \\ y' &= y^2 - x^2. \end{aligned}$$

(a) Sketch the nullclines, trajectory directions and equilibria.

(b) Find the linearization at one equilibrium and classify it.

Chapter 3

Modeling Population

3.1 Introduction

"Excessive (population) growth may reduce output per worker, repress levels of living for the masses and engender strife."
  - Confucius

The growth and limits to growth of human populations is a question of great importance. Planning in health care, transportation, education and so on requires projections of populations whose needs are to be addressed. Harkening back to Chapter 1, population growth is also related to the intensity of wars. The data analysis of Steffler 1974 indicated that:

"for each interval in which human population doubles,
there occurs wars in which ten times as many people
are killed as in the previous interval."

Thus, one fundamental problem is to make reliable projections of future populations. One of the first attempts is due to Thomas Malthus. The Malthusian model gives a good estimate of growth in some cases and over small times but is quite wrong over longer times. The leading correction was developed by Quetelet and Verhulst. It was rediscovered and tested against data by Reed and Pearl. We develop these models in this chapter. These two models are applied to U.S. census data between 1790 and 1990.

Important figures in the understanding of how human populations grow include:

T.R. Malthus- 1766-1834 - was a interesting figure. He was both a mathematician and a minister with a dour outlook on the possibilities of humanity. He believed, for example, that

"social progress was illusory".

Malthus asserted that food increases linearly[1] while population geometrically

$$N'(t) = aN(t) \text{ where } a > 0. \qquad \text{(Malthus' model)}$$

His conclusion was that

> *"the superior power of population*
> *is resolvable only by misery."*

His essay is short, tightly reasoned and still interesting to read. He is also credited with founding the field of Economics as a quantitative science.

Lambert Adolphe Jacques Quetelet - 1796 – 1874- was an astronomer, mathematician, statistician and sociologist who believed improvement of the condition of humankind was possible:

> *"As laws and the principles of religion and morality are influencing causes, I have then not only the hope, but the positive conviction, that society may be ameliorated and reformed."*
>
> *-L. Quetelet*

Quetelet reasoned by analogy to motion through a resistive medium (in which friction is proportional to *velocity*^{2}) that population should grow with a quadratic correction (proportional to *population*^{2}) to Malthus' model by

$$N' = \alpha N - \beta N^2, \text{ where } \beta << \alpha.$$

His work was the first on the logistic model.

P.F. Verhulst - 1804–1849- argued, following ideas of Quetelet, that Malthus' model was limited in applicability:

> *"We shall not insist on the hypothesis of geometric progression, given that it can hold only in very special circumstances; for example, when a fertile territory of almost unlimited size happens to be inhabited by people with an advanced civilization, as was the case for the first American colonies."*
>
> *-P.F. Verhulst*

He fully developed the logistic model for growth. However, his work was largely ignored because he did not attempt to test it against data, being hindered by a lack of good data. Verhulst's contribution is now increasingly recognized[2]. Based on his logistic model, in 1845 Verhulst predicted a maximum population of several European countries including

Belgium: $6,600,000$ (far lower than today),

France: $40,000,000$ (a fairly accurate prediction).

Verhulst's population predictions

---

[1] Malthus has an extensive justification for this assumption in his writing. His basic idea is that the supply of arable land is limited and it takes more effort to grow crops as less productive land is cultivated. Some have suggested (long after Malthus) that his assertion was based on agricultural production data from his era. Others have suggested that his mathematical tools for studying growth were only arithmetic progressions and geometric progressions so agricultural data was forced to fit the former.

[2] Nevertheless, mathematicians modeling a phenomena should take from his case the importance of connecting theories to data.

R. Pearl and L.J. Reed -The logistic model was rediscovered by Pearl and Reed in their study of the US population. The accuracy of its predictions for the US population was the key to immediate acceptance of the model.

3.2   Malthus' exponential model

"Consider... the nature of those checks which have been classed under the general heads of Preventive and Positive.

It will be found that they are all resolvable into moral restraint, vice and misery."

- Thomas Malthus, from his article on Population Control in the 1815 Britannica.

To develop the Malthusian model, it is convenient to consider the number of members of a large population, $N(t)$, as a real valued function rather than integer valued. Malthus postulated that

> *"First, That food is necessary to the existence of man.*
> *Secondly, That the passion between the sexes is necessary*
> *and will remain nearly in its present state."*

From these postulates he derived the law that

> *"Population, when unchecked, grows in a geometric ratio.*
> *Subsistence increases only in an arithmetic ratio."*

If the members of the population reproduce at a constant rate for all individuals for all time and there is no immigration then N must satisfy the equation describing a pure birth process:

$$N'(t) = bN(t),$$

where $b > 0$ is defined to be the birth rate constant. If there are only deaths with no births or immigration then $N(t)$ satisfies the equations describing a pure death process:

$$N'(t) = -dN(t),$$

where $d > 0$ is defined as the death rate. Generally $b > d$ so that setting $a = b - d > 0$, a basic model of the combined effects of births and deaths is:

$$N'(t) = aN(t), \quad N(0) = N_0 \text{ given.} \tag{3.1}$$

Mathematically, birth and death and two comparable processes. In populations they are distinct. One historical avenue for population growth is as follows. A population is in balance ($b = d$) then some improvement in public sanitation or heath occurs and the death rate drops so $a = b - d > 0$. Exponential growth occurs and after 10-20 years, the birth rate drops and the population re-stabilizes at a higher level.

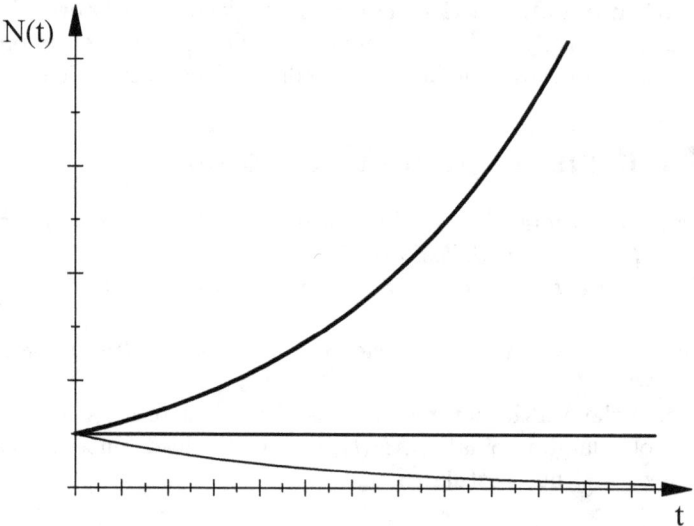

Figure 3.1:   Population N(t) predicted by Malthus' model:
a>0 (exponential growth), a=0 (steady) and a<0 (exponential decay)

The solution of this equation is well known: $N(t) = N_0 e^{at}$. Solutions $N(t)$ in the three cases of $a > 0, a = 0, a < 0$ are plotted in Figure 3.1 below.If the population is known at two different times, $N(t_1)$ and $N(t_2)$, then the rate constant $a = b - d$ is calculated by the formula

$$a = \frac{\ln N(t_1) - ln N(t_2)}{t_1 - t_2}.$$

The most common way to represent data of growth rates is not by giving the constant a but rather in terms of

$$growth\ rate\ =\ (live\ births\text{-}deaths)\ per\ thousand,\ per\ year$$
$$= \frac{N(t+1) - N(t)}{N(t)},\ if\ \ t = years\ \ and\ \ N(t) = population(thousands).$$

This tabulated *growth rate* is not the same as the rate constant a in equation (3.1). (For a small the *growth rate* is an approximation to the rate constant a in the equation however.) It is an interesting exercise to take the following countries' growth rates and compute their rate constants a and population doubling times.

Germany	Denmark	US & Japan	Argentina	Afghanistan	Costa-Rica	Kuwait
0.1	0.5	1.1	1.5	2.5	4	8

Growth rates: relative population change per year

Population grows geometrically. The model (3.1) makes several interesting predictions. First, the population at successive years

$$N(0), N(1), N(2), N(3), \cdots$$

does indeed form a geometric ratio (or sequence), as asserted by Malthus.

Definition. *A sequence* $a_1, a_2, a_3, a_4, \cdots$ *is geometric if there are fixed numbers* c, r *such that*

$$a_n = cr^n \text{ for all } n.$$

A geometric sequence with $r > 1$ is a discrete time equivalent to exponential growth. To verify his assertion that the populations $N(0), N(1), N(2), \cdots$ form a geometric sequence, we use the geometric sequence test.

Theorem. [Geometric Sequence Test] *A sequence* $a_1, a_2, a_3, a_4, \cdots$ *is geometric if the ratio of successive terms is a fixed constant, independent of* n:

$$\frac{a_{n+1}}{a_n} \equiv \alpha, \text{ for all } n.$$

Applying the test, from $N(t) = N_0 e^{at}$, it follows that

$$\frac{N(t+1)}{N(t)} = \frac{N_0 e^{a(t+1)}}{N_0 e^{at}} = e^a, \quad \text{for all times } t.$$

Doubling times. The model also predicts a fixed doubling time (if $a > 0$) or half life (if $a < 0$). Indeed, let $T = $ *doubling time*. Then

$$
\begin{aligned}
N(t+T) &= 2N(t) \\
&\Leftrightarrow \\
N_0 e^{a(t+T)} &= 2N_0 e^{at}.
\end{aligned}
$$

We then have

$$2 = \frac{N(t+T)}{N(t)} = \frac{N_0 e^{a(t+T)}}{N_0 e^{at}} = e^{aT}.$$

Solving for the doubling time T gives:

$$T = \frac{\ln 2}{a}.$$

As an example the following doubling times for the world population are cited in the 1995 book of Cohen.

Approx. Date	Description	Pop. billions	Doubling years Before	Doubling years After
8000 BC	local agriculture	0.005	40,000-300,000	1,400-3,000
1750	global agriculture	0.75	750-1800	100-130
1950	public health	2.5	87	36
1970	fertility	3.7	34	40

Demographers quantify population growth by the relative change and the percent change every year.

Definition 24 *The relative change over one time unit is*

$$Relative\ change := \frac{N(t+1) - N(t)}{N(t)}$$

and the percent change is

$$Percent\ change = Relative\ change \times 100 = \frac{N(t+1) - N(t)}{N(t)} \times 100.$$

For solutions of the Malthusian model, the relative change and percent change are also fixed:

$$
\begin{aligned}
Relative\ change &= \frac{N(t+1) - N(t)}{N(t)} \\
&= \frac{N_0 e^{a(t+1)} - N_0 e^{at}}{N_0 e^{at}} \\
&= e^a - 1, \text{ for all time } t, \\
&\text{and} \\
Percent\ change &= 100 \times (e^a - 1), \text{ for all time } t.
\end{aligned}
$$

Consequences of Malthus' Model. The Malthusian law is so often presented as a mathematical toy for exploring exponential growth that it is easy to forget that it has serious consequences for the structure of society, once its hidden assumptions are accepted. Alfred North Whitehead wrote the following in his 1933 book *The Adventure of Ideas*.

"Another instance of ill-judged simplification is the use made of the Malthusian Law of Population. ... It was then concluded, as a consequence of this Malthusian Law, that population will always overtake the means of subsistence. ... , the normal structure of society was that of a comparatively affluent minority subsisting on the labours of a teeming population checked by starvation ...there must be a pool of labour, starving and destitute, ready to work on the wages of bare subsistence. Factories taking advantage of such cheap labour will drive out of trade those managed on fanciful humanitarian lines." - A.N. Whitehead.

Application: modeling the growth of the U.S. population

Forecasting population levels accurately is of profound importance for society and governments. As an example, consider the following analysis of Abraham Lincoln:

"Taking the nation in the aggregate, and we find its population and ratio of increase for the several decennial periods to be as follows:

Year - Population - Ratio of increase.

- - Per cent.

1790 - 3,929,827 - ....

1800 - 5,304,937 - 35.02

1810 - 7,239,814 - 36.45

1820 - 9,638,131 - 36.45

1830 - 12,866,020 - 33.49

1840 - 17,069,453 - 32.67

1850 - 23,191,876 - 35.87

1860 - 31,443,790 - 35.58

This shows an average decennial increase of 34.60 per cent in population through the seventy years from our first to our last census yet taken. It is seen that the ratio of increase at no one of these seven periods is either 2 per cent below or 2 per cent above the average, thus showing how inflexible, and consequently how reliable, the law of increase in our case is. Assuming that it will continue, it gives the following results:

Year - Population

1870 - 42,323,341

1880 - 56,967,216

1890 - 76,677,872

1900 - 103,208,415

1910 - 138,918,526

1920 - 186,984,335

1930 - 251,680,914

· · · if we do not ourselves relinquish the chance by the folly and evils of disunion or by long and exhausting war springing from the only great element of national discord among us."

– Abraham Lincoln, State of the Union Address, December 1862.

Abraham Lincoln recounted sitting on a riverbank and teaching himself mathematics from a borrowed copy of Euclid's Elements. From his above address, he also mastered modeling population growth and made policy based on his analysis. This analysis is based on census data, available for the American population every ten years, beginning with 1790. Consider the first few data points.

Year	1790	1800	1810	1820	1830	1840	1850	1860
Pop.	3.9	5.3	7.2	9.6	12.9	17.1	23.2	31.4

American census data (in millions)

There are two parameters in the exponential growth model: the initial population $N_0 = N(0)$ and the growth rate constant

$$a = \frac{N'(t)}{N(t)}.$$

Using the population in 1790 and 1800 we calculate the growth rate constant by

$$a = \frac{1}{10} \ln \left(\frac{N(1800)}{N(1790)} \right) = 0.3067.$$

This value corresponds to a 35.9% increase[3] every 10 years:

$$
\begin{aligned}
N(t+10) &= 1.359 N(t) \\
&\text{and} \\
N(1790+t) &= N(1790) e^{0.3067t}
\end{aligned}
$$

This formula has remarkable predictive power for the period 1790 to 1860, see Table 1.4.

Year	1790	1800	1810	1820	1830	1840	1850	1860
Actual	3.9	5.3	7.2	9.6	12.9	17.1	23.2	31.4
Predict	3.9	5.3	7.1	9.5	12.8	17.3	23.3	31.4

Predicted versus Actual U.S. Population, 1790-1860,

Exponential model, Population in Millions.

During this period the U.S. was thinly populated so the carrying capacity of the U.S. was far above the actual population. Thus, it should not be surprising that the exponential model is reasonable. Still, the fit is much better than reasonable, it is remarkable!

3.2.1 The error in the growth rate

"fallor ergo sum." - Augustine.

"As far as the laws of mathematics refer to reality, they are not certain; and as far as they are certain, they do not refer to reality." - Albert Einstein

As a step to assessing the error in prediction, we first consider the error in estimating the growth rate. As error is an essential feature, there are trademarked[4] terms.

Definition 25 *The error, relative error and % error are, respectively,*

$$
\begin{aligned}
error &= \textit{True Value - Approximate Value} \\
relative\ error &= \left| \frac{\textit{True - Approximate}}{\textit{True}} \right| \\
\%\ error &= \textit{Relative error} \times 100.
\end{aligned}
$$

---

[3] This is very close to the value 34.6% calculated by Lincoln.

[4] "Standard" terms is another description but "trademarked" makes it clear that their definitions are set, making communication with others possible.

Recall the census data for 1790, 1800 and 1810

Year	US Population
1790	$3,929,214$
1800	$5,308,483$
1810	$7,239,881$

During these years the US population was undergoing nearly perfect exponential growth with rate constant $a = 0.3067$. Alternately, the growth rate constant $a = N'/N$ can be *estimated* by approximating N', N at intermediate years using differences for N' and averages for N. Richardson's idea from Chapter 1 then gives an estimated value of

$$a = \frac{N'(t)}{N(t)} \simeq \frac{\frac{N(t+10)-N(t)}{10}}{\frac{N(t+10)+N(t)}{2}}.$$

This is exact if "10" is replaced by Δt and $\Delta t \to 0$. Since "10" is not small it is only an estimate. We calculate the approximation

$$
\begin{aligned}
a_{approximate} &\simeq \frac{\frac{N(1800)-N(1790)}{10}}{\frac{N(1790)+N(1800)}{2}} \\
&= 0.5 \times \frac{5308483 - 3929214}{5308483 + 3929214} \\
&= 7.4654 \times 10^{-2}.
\end{aligned}
$$

The value that is *as exact as the model and data allows* is

$$a_{exact} = \frac{\ln\left(\frac{5308483}{3929214}\right)}{10.0} = 3.0087 \times 10^{-2}.$$

The relative error in the approximation 0.0746 is 1.5 (i.e., 150% error):

$$\text{relative error} = \left| \frac{3.0087 \times 10^{-2} - 7.4654 \times 10^{-2}}{3.0087 \times 10^{-2}} \right| = 1.5.$$

Thus, since $\Delta t >> 1$ this approximation gives an reasonable estimate of the parameter magnitude but an inaccurate value.

3.3 The logistic Model

"At the same ratio of increase which we have maintained, on an average, from our first national census, in 1790, until that of 1860, we should in 1900 have a population of 103,208,415. And why may we not continue that ratio far beyond that period? Our abundant room, our broad national homestead, is our ample resource. Were our territory as limited as are the British Isles, very certainly our population could not expand as stated. "

– Abraham Lincoln, State of the Union Address, December 1862.

"The good news is world population growth rate decreases systematically and is expected to reach zero by 2050, thanks to urbanization and women's education." - Dan Shechtman

Whitehead had several criticisms of the Malthusian Law of population including:

" ... in truth there is a complex situation depending on the balance of many factors. By arbitrarily seizing upon one or two factors ... almost any law of population can be deduced. Thus the Malthusian Law, with its sociological consequences, is not an iron necessity. It is a possibility ... during the eleven centuries since the age of Charlemagne to the present day a persistently increasing population has been accompanied by an equally persistent rise in the general standard of life." - p. 81 in: A.N. Whitehead, The Adventure of Ideas, 1933.

Among these many possibilities is that exponential growth is inherently self limiting and cannot continue on indefinitely. One fundamental question therefore is, other than by misery,

How does a system which begins growing exponentially limit itself?

To study this, we suppose that the environment of the population has a fixed maximum carrying capacity K. We assume that at "low" population density the growth rate is r (exponential growth). As the population density increases, the growth rate decreases. Specifically, as the population approaches the carrying capacity, K, the growth rate approaches zero. A straightforward way to model this growth rate is with the linear function shown below. Inserting this relation in for the rate constants

$$a = r \left(1 - \frac{N}{K} \right)$$

in equation 3.1 yields the famous logistic equation of Verhulst and Pearl:

$$N'(t) = r \left(1 - \frac{N(t)}{K} \right) N(t) \tag{3.2}$$

One quantity of interest is the carrying capacity K. This leads to the questions:

What is K?
Is K a coefficient in a differential equation or a fundamental property of a population?

Calling it the carrying capacity does not answer the question and the question becomes more unclear the deeper one looks. Cohen [1995] in Table 2.1 gives 8 definitions (a few are summarized below) for a populations carrying capacity, some ecological, some circular and some stating simply that K is a parameter

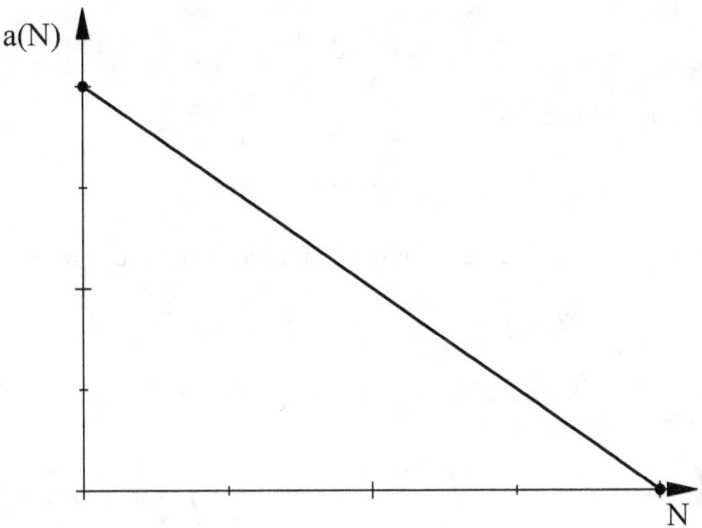

Figure 3.2: Birth rate $a(N) = r(1 - N/K)$ decreases as $N \to K$

in the logistic equation.

" ... number at which these curves (birth and death rates) cross... idealized concept not to be taken literally ..." " number ... area can support indefinitely" "asymptote of the sigmoidal curve..."	"maximum population size an area can support without reducing its ability to support the same species in the future" "number ... a habitat can support." " size ... approached in time..."

Definitions of K

3.3.1 Solving the logistic equation

"I consider that I understand an equation when I can predict the properties of its solutions, without actually solving it." - Paul A. M. Dirac

The logistic equation

$$N' = r \left(1 - \frac{N}{K}\right) N$$

can be solved by separation of variables and doing the integrals that arise by a standard partial fractions decomposition. Some care is required to avoid taking the logarithm of a negative number. We do it in two cases.

Case 1: Increasing populations N(0) < K. In this case $N(t)$ is increasing since

$$1 - \frac{N}{K} \geq 0.$$

Starting with

$$N' = r\left(1 - \frac{N}{K}\right)N,$$

Separating variables gives

$$\int \frac{dN}{N(1 - N/K)} = \int r\,dt.$$

The RHS is $rt + C$. The LHS requires partial fractions to integrate[5]. We have

$$\int r\,dt = \int \frac{dN}{N(1 - N/K)} = \int \frac{1}{N} + \frac{1}{K - N}\,dN.$$

In this case $N > 0$ and $K - N > 0$ so we can just integrate. This gives $\int \frac{1}{N}dN = \ln(N) + C$ and

$$\begin{aligned}\int \frac{1}{K - N}dN &= -\int \frac{1}{K - N}(-1\,dN) = -\int \frac{1}{u}du, u = K - N \\ &= -\ln(u) + C \\ &= -\ln(K - N) + C\end{aligned}$$

Then we have

$$\ln(N) - \ln(K - N) = rt + C$$
$$or$$
$$\ln\left(\frac{N}{K - N}\right) = rt + C, \text{ when } N(0) < K.$$

Thus, we have $N/(K - N) = \exp(rt + C)$. Solving for N,

$$N(t) = \frac{K}{1 + \exp(-[rt + C])}.$$

This solution when $N(0) < K$ is plotted below in Figure 3.3.

Case 2: Decreasing populations N(0) > K.

"In demographic terms, Europe is vanishing. Twenty or so years from now our countries will be empty."

- Jacques Chirac in 1984[6].

---

[5]Starting with
$$\frac{1}{N(1 - N/K)} = \frac{A}{N} + \frac{B}{1 - N/K}$$
and solving for A, B in the usual way (get common denominator, equate like powers etc) we have
$$\frac{1}{N(1 - N/K)} = \frac{1}{N} + \frac{(1/K)}{1 - N/K} = \frac{1}{N} + \frac{1}{K - N}$$

[6]This quote illustrates the danger of predictions that can be checked within ones lifetime. Sadly, it also illustrates the fact that population predictions are linked to racial politics because here Chirac is talking about Europe being empty of people he regards as the 'right kind' of Europeans.

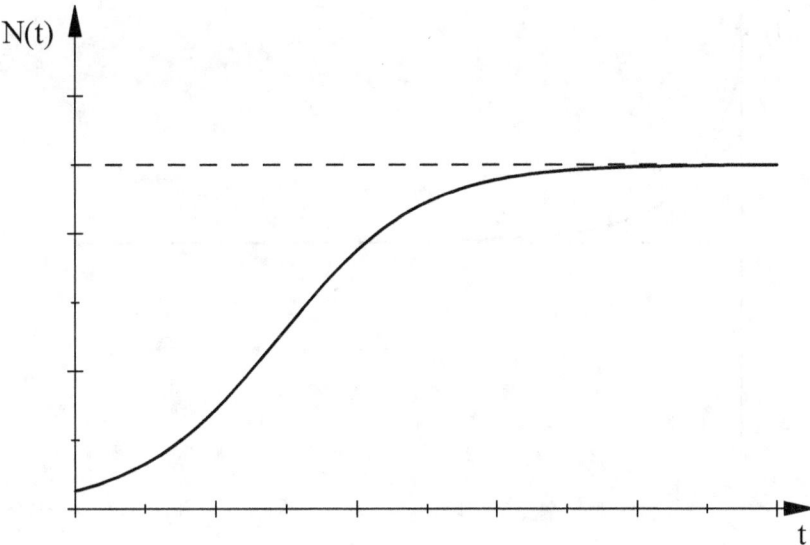

Figure 3.3: The solution when $N(0) < K$
Observe $N(t) \to K$ as $t \to \infty$.

This is the case when the population is declining. When $N(0) > K$, $N(t)$ decreases to K as t increases. When $N(0) > K$ the analysis needs to be modified to be careful about signs as follow. As before we separate variables. This gives

$$\int \frac{1}{N} + \frac{1}{K-N} dN = rt + C.$$

Since $N > 0, 1/N > 0$ and

$$\int \frac{1}{N} dN = \ln(N) + C.$$

However $N(0) > K$ so $K - N < 0$. The $K - N$ term has to be rewritten to be positive by

$$\int \frac{1}{N} dN - \int \frac{1}{N-K} dN = rt + C.$$

Then we have

$$\ln(N) - \ln(N - K) = rt + C, \text{ or } \ln(\frac{N}{N-K}) = rt + C \text{ when } N(0) > K.$$

Solving for $N(t)$ gives

$$N(t) = \frac{K}{1 - \exp(-[rt + C])}, \text{ where } C = \ln\left(\frac{N(0)}{N(0) - K}\right)$$

The solution in this case when $N(0) > K$ is plotted next in Figure 3.4.

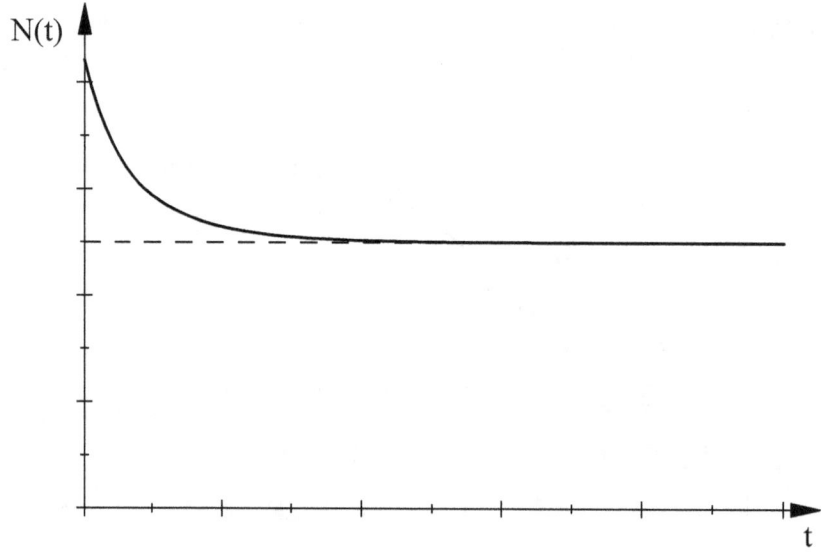

Figure 3.4:   The solution when $N(0) > K$:
$N(t)$ decreases to K as $t \to \infty$.

3.3.2   Three Graphs of logistic Model

The solution to the logistic equation is given by

$$
N(t) = \begin{cases} \frac{K}{1 + \exp(-[rt+C])}, & \text{when } N(0) < K, \\ & \& \\ \frac{K}{1 - \exp(-[rt+C])}, & \text{when } N(0) > K, \end{cases} \tag{3.3}
$$

where C is a constant of integration. When $N(0) < K$, the solution $N(t)$ to equation (3.2) has the $\mathcal{S}$ shaped profile depicted below. The predicted population from the logistics equation is sometimes called the $\mathcal{S}$ curve due to its shape in Figure 3.5 next.   The second plot of importance is the plot of the growth rate $a(N)$ against N, Figure 3.6. If the logistic model is accurate, this is a line. Data plotted $N'/N, N$ gives a quick check as to whether the logistic model is a good description of the phenomena being studied.   If the data is a reasonable fit then the model parameters can be found from this plot by a linear least squares and then reading off the plot

$$K \quad = \quad \text{horizontal intercept,}$$
$$\text{and}$$
$$r \quad = \quad \text{vertical intercept.}$$

The third plot of importance is $N'(t)$ against t. This is important because for many applications the fundamental data is the change over the last time

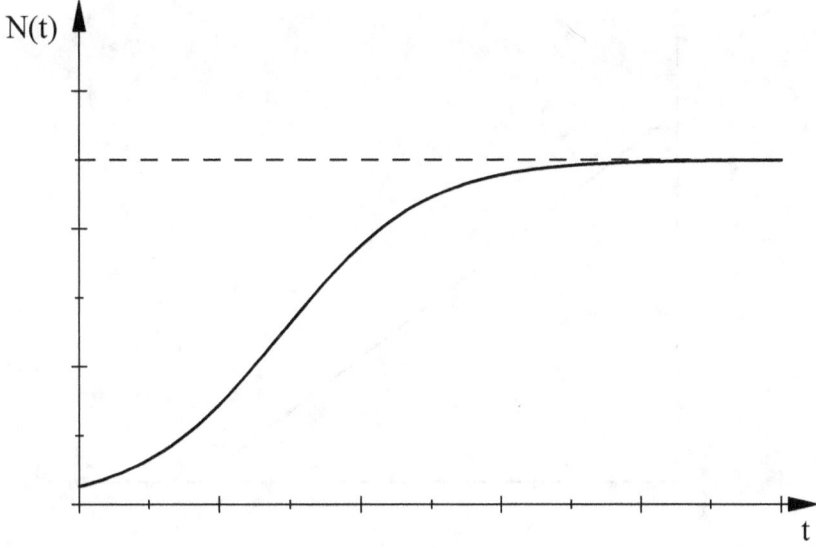

Figure 3.5:   The Logistic or $\mathcal{S}$-curve: when $N(0) < K$ $N(t)$ first grows exponentially then slows and $N(t) \to K$ as $t \to \infty$.

period rather than the (possibly unknowable) aggregate total $N(t)$. This plot is known as *Hubbert's pimple* for its shape and for his use of it to forecast peak oil production.

3.4   Calibration of the logistic Model

The equation (3.2) (or, equivalently, the solution $N(t)$ given by (3.3)) has 3 parameters r, K and $N(0)$ (equivalently, r, K and C). These can be determined when 3 population levels are observed. Indeed, equation (3.3) can be written as

$$C - rt = \ln \frac{K - N}{N}$$

so that we obtain the following system of equations from observations at three different times, $(t_0, N_0), (t_1, N_1)$ and (t_2, N_2):

$$\begin{cases} C - rt_0 = \ln \frac{K - N_0}{N_0}, \\ C - rt_1 = \ln \frac{K - N_1}{N_1}, \\ C - rt_2 = \ln \frac{K - N_2}{N_2}. \end{cases} \tag{3.4}$$

For general data this is a system of 3 nonlinear equations in 3 variables which is solved for the constants C, r and K using a numerical procedure like Newton's

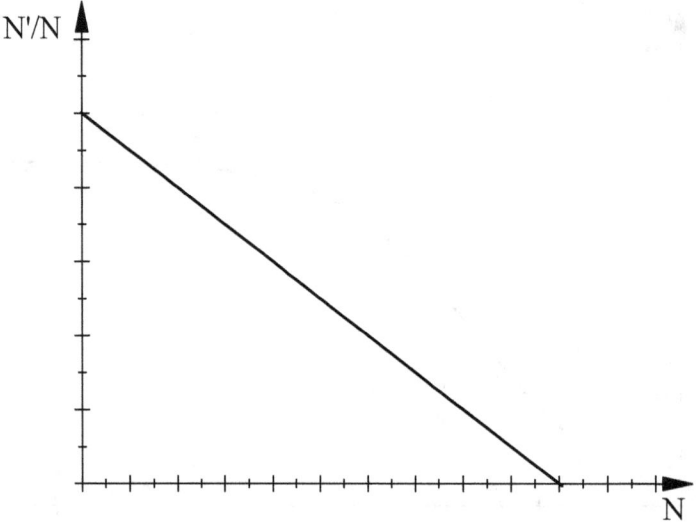

Figure 3.6: Logistic growth rate: N'/N vs. N

method. For equally spaced data an explicit formula for K is possible and derived next.

Special case: equally spaced data

Suppose $t_0 < t_1 < t_2$ are equally spaced time intervals,

$$\triangle t = t_2 - t_1 = t_1 - t_0.$$

Then, the system (3.4) has the solution:

$$K = \frac{N_1(2N_0 N_2 - N_0 N_1 - N_1 N_2)}{N_0 N_2 - N_1^2} \tag{3.5}$$

Once K is determined, any two equations from the system (3.4) reduce to a 2×2 linear system for C and r.

   Derivation of the formula for K. *The derivation of the formula (3.5) is by algebraic manipulation[7] of the equations*

   Solving for C in each equation we have

$$C = rt_j - \ln \frac{K - N_j}{N_j}.$$

---

[7] The general way nonlinear equations are solved exactly is as follows: select one variable in one equation. Solve for that variable in terms of the others and substitute. This reduces the problem to one fewer equation and one fewer variable. Continue.

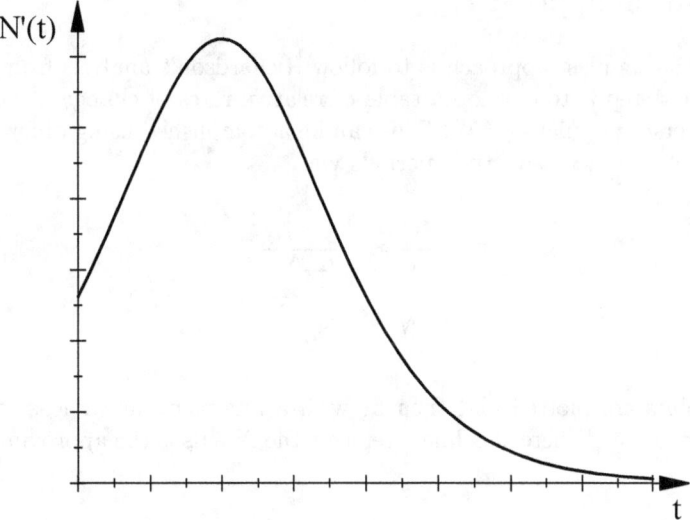

Figure 3.7:
Hubbert's pimple: $N'(t)$ vs. t
For the oil industry, N' is the yearly production

Eliminating C using the above we have the following system of 2 nonlinear equations for 2 variables:

$$r(t_1 - t_0) = \ln\left(\frac{N_0}{N_1}\frac{K - N_1}{K - N_0}\right)$$

$$r(t_2 - t_1) = \ln\left(\frac{N_1}{N_2}\frac{K - N_2}{K - N_1}\right)$$

Next eliminate r by dividing the equations. This gives one nonlinear equation for 1 unknown, K,

$$\frac{t_1 - t_0}{t_2 - t_1} = \frac{\ln\left(\frac{N_0}{N_1}\frac{K-N_1}{K-N_0}\right)}{\ln\left(\frac{N_0}{N_1}\frac{K-N_1}{K-N_0}\right)}.$$

If the data is equi-spaced then $\frac{t_1 - t_0}{t_2 - t_1} = 1$:

$$\ln\left(\frac{N_0}{N_1}\frac{K - N_1}{K - N_0}\right) = \ln\left(\frac{N_0}{N_1}\frac{K - N_1}{K - N_0}\right)$$

$$or$$

$$\frac{N_0}{N_1}\frac{K - N_1}{K - N_0} = \frac{N_0}{N_1}\frac{K - N_1}{K - N_0}.$$

Cross multiplying, solving using the quadratic formula and rearranging gives the claimed formula for K.

A graphical approach

Perhaps the simplest approach is to follow Richardson's analysis from Chapter 1. The first step is to develop a table of relative rates of change in population (N'/N) versus population (N). This can be accomplished using differences and averages of two successive time periods via,

$$\frac{N'}{N} \approx \frac{N_2 - N_1}{\frac{N_1 + N_2}{2}}$$
$$N \approx \frac{N_1 + N_2}{2}$$

These points are plotted and then fit with a line using linear least squares / linear regression. Where this line intersects the N axis is the approximate value of K.

3.5   The U.S. Population

One of the principal objects of theoretical research in my department of knowledge is to find the point of view from which the subject appears in its greatest simplicity.

- Gibbs, Josiah Willard (1839 - 1903)

In principle, any three year' census data can be used to find C, K and r. However, if 1790, 1800 and 1810 are used difficulties arise because that is a period of nearly pure exponential growth. Another way to view it is that the influence of the non exponential term involving N_2 in the logistic equation (1.4) is, during these years, smaller than the accuracy to which the population figures are known. Either more accurate census data is needed (not possible in 1790) or data from years in which the logistic correction term involving N^2 is significant is needed. If the data from 1790, 1850 and 1910 is used to determine C, r, and K we arrive at a carrying capacity of the U.S. of

$$\text{carrying capacity } K = 197.274 \text{ million,}$$

and a closed form formula for the predicted U.S. Population given by:

$$N(t) = \frac{197.274}{1 + e^{3.896 - 0.031t}} \text{ millions.}$$

These are precisely the dates and data used in the 1924 study of the U.S. Population done by Raymond Pearl and Lowell J. Reid. Their analysis produced a remarkably good fit to the U.S. population from 1790 to 1950. See Table??)

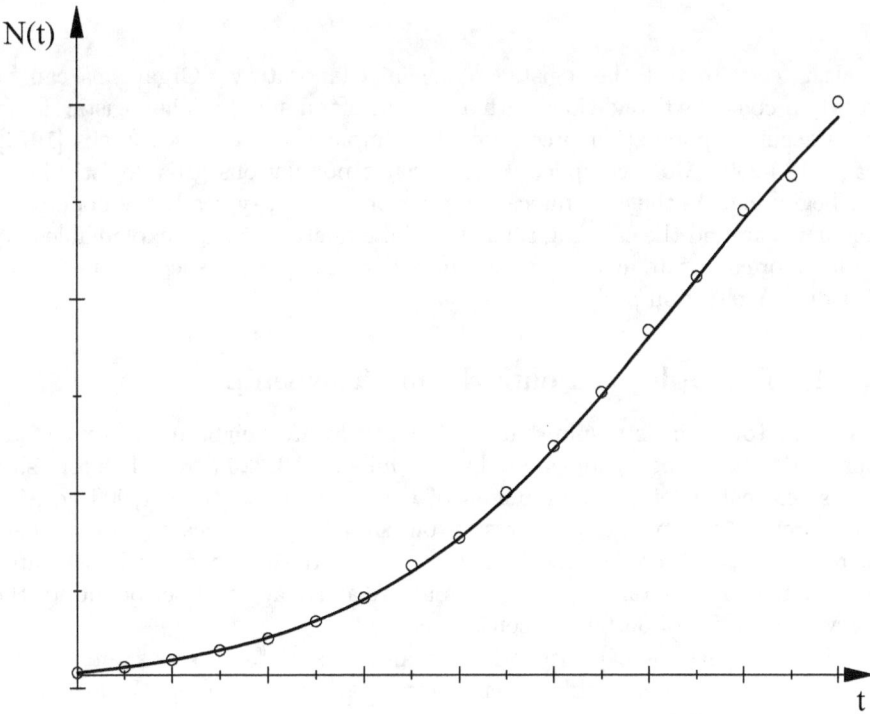

Figure 3.8: Logistic Model (curve) and actual US population (dots): 1790 to 1950.

or the plot of the true and values predicted by the logistic model in Figure 3.8.

Year	Actual Population	Exponential Model	Logistic Model
1790	3.93	3.93	3.93
1800	5.31	5.29	5.34
1810	7.24	7.11	7.23
1820	9.64	9.56	9.76
1830	12.87	12.87	13.11
1840	17.07	17.31	17.51
1850	23.19	23.28	23.19
1860	31.44	31.32	30.41
1870	38.56	42.13	39.37
1880	50.16	56.68	50.18
1890	62.95	76.24	62.77
1900	76.00	102.56	76.87
1910	91.97	137.96	91.97
1920	105.71	185.59	107.40
1930	122.78	249.66	122.40
1940	131.67	335.84	136.32
1950	150.70	451.78	148.68

Table: Predicted vs. Actual US populations

It is easy to test the logistic model in a laboratory. Organisms can be grown in controlled conditions with a constant food supply. The logistic model gives accurate population predictions for simple organisms, see Krebs [1972], pages 190-200. More complicated organism's populations grow logistically in the beginning. As the carrying capacity is approached, typically the population fluctuates around the carrying capacity. These fluctuations are explainable only with a more accurate model that accounts, for example, for such effects as time lags due to gestation periods.

3.5.1   Example: A Pennsylvania Township

Hampton township is a suburban community located eight miles from Pittsburgh, PA. It is roughly four miles by four miles $= 10,323$ acres. The township limits residential lots to a minimum of one acre. It contains $1,000$ acres of commercial property and 400 acres of township properties leaving a maximum of roughly $9,000$ acres of residential property. If this property is saturated with homes each containing family of four, we arrive at an upper bound for the carrying capacity of $36,000$ persons.

However, carrying capacity is limited also by such factors as sanitary facilities, infrastructure, available water, transportation, and so on. We can attain a, hopefully more realistic, estimate using the census data:

Year	Population
1970	12515
1980	14260
1990	15568

Census: Hampton Township

Using this equally spaced census data, and assuming the logistic equation

$$N' = r \left(1 - \frac{N}{K} \right) N$$

holds, we can use equation (3.5) to give an estimate for K, as follows. Inserting the values

$$N_0 = 12515, \quad N_1 = 14260, \quad N_2 = 15568$$

from the census data gives an estimate of K:

$$K = 18083.$$

This value is roughly half the size of $36,000$ and certainly more realistic. A better estimate of the school age population would follow by considering the age structure of the overall population of Hampton township.

3.6 Other models of a single population

The logistic model of population growth is rather crude; it contains only two parameters and one initial condition. Nevertheless, it can have remarkable predictive ability. Reed and Pearl established its remarkable accuracy for the US population for the years 1790-1990.

If the logistic model is refined, it is typically coupled with other factors which influence the levels and growth of human populations including:

- Immigration,

- Medical advances increasing life expectancy,

- Changes in age structure of the population,

- Economic changes, and

- Ethnic or religious subgroups with differing birth rates.

These factors are discussed, for example, in Pollard [1973]. For other non-human populations various modifications of the logistic model's growth rate parameter $a(N)$ have been proposed:

1. **Harvesting or migration at a constant rate.**
   This adds an additional harvesting term, $-hN$ to the logistic equation where the constant h represents the harvesting rate:

   $$N' = rN\left(1 - \frac{N}{K}\right) - hN.$$

2. **Per capita growth increases with N for small values of N.**
   This occurs for schools of certain fish and other groups of animals. In this case the function $a(N)$ resembles the following figure.

3. **Negative growth rate if population levels are too small.**
   This can be caused by dispersion of the population over too large an area or lack of genetic diversity. In this case the function $a(N)$ resembles the next figure. This profile is also interesting for forecasting social trends. For example, if you see a few people wearing their clothes backwards, the normal thought would be that those people are odd. If lots of people do it, including sports and music stars, suddenly it starts to look cool and people emulate the trend. When too many people do it, it becomes odd again.

4. **Growth rates decreasing as population increases.** Examples include:

   - Gompertz, 1825 proposed $a(N) = r \log\left(\frac{K}{N}\right)$:
   - F. Smith, 1963 proposed $a(N) = \frac{r(K-N)}{K+\alpha N}$:

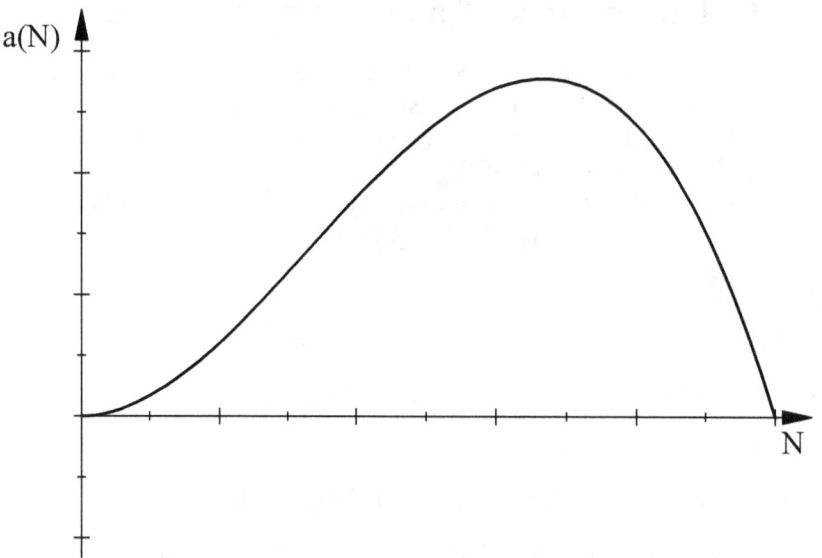

Figure 3.9: Growth rate a(N) increases for small N

- Ayala, Gilpin and Ehrenfeld, 1973 proposed $a(N) = r\left(1 - \left(\frac{N}{K}\right)^{\theta}\right)$:

- Nisbeth and Gurney, 1982 proposed $a(N) = re^{1 - N/K} - d$:

5. **Fluctuating carrying capacities due to seasonal or other changes.**
   This leads to a logistic equation with K being a function of t.

   $$N' = rN\left(1 - \frac{N}{K(t)}\right),$$

   where $K(t)$ is a fluctuating carrying capacity.

6. **Age structure models.**
   For human populations, the age structure of the population plays a critical role in forecasting future population levels. Age structure is the next logical refinement to introduce in a human population model.

7. **Populations with exotic mating patterns such as trigger fish.**
   The life cycle of trigger fish is interesting. Trigger fish spend the first half of their life as females and then change to males for the second half of their life.

Conclusion 26 *Looking at the diversity of growth models, our conclusion is now simple: Do not blindly do a linear regression to determine a logistic model. Plot the data on axes $\left(\frac{N'}{N}, N\right)$ and look at what the data suggests the growth*

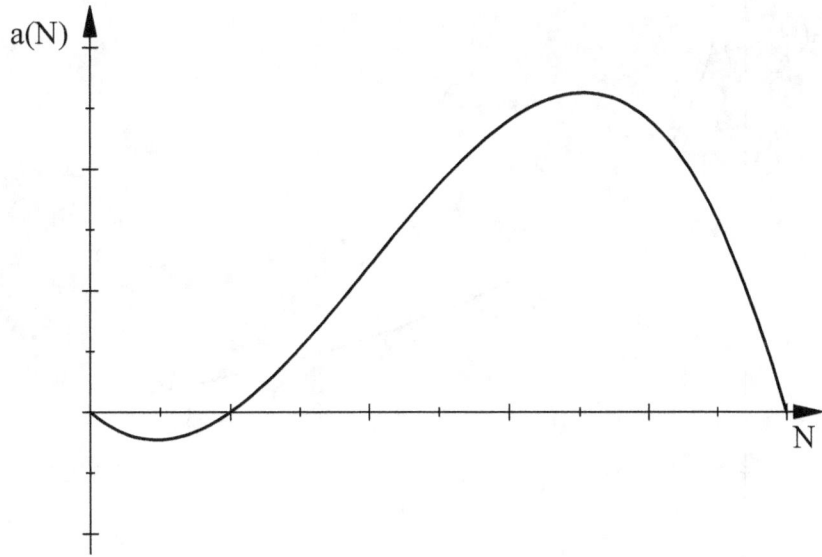

Figure 3.10: a(N) negative for small N

rate $a(N)$ should be. Then determine the specific $a(N)$ by doing data fit among functions with those shapes.

3.7 Reproduction at discrete time intervals

"... discrete mathematics is more difficult than continuous mathematics. If you look at formulas for derivatives of reciprocals and then finite differences for reciprocals, you see how things are more complicated in the discrete case. ... The main point in the theory of difference approximations is to prove stability. To prove stability is like getting an a priori estimate for the solution of the equation. But to get those estimates for difference approximations is much more sophisticated than to get them for a differential equation."

- Peter Lax, MAA Focus (May/June 2005).

Many prey species reproduce at discrete time intervals and all at the same time with non-overlapping generations[8]. Thus, the basic assumption that $N(t)$ is continuous (and differentiable) is violated and the correct model is a difference equation. We now turn to the discrete time, logistic difference equation, known as the Ricker model. The Ricker model has proven useful in improving efficiency

---

[8] With overlapping generations the birth process is discrete but the death process operates continuously.

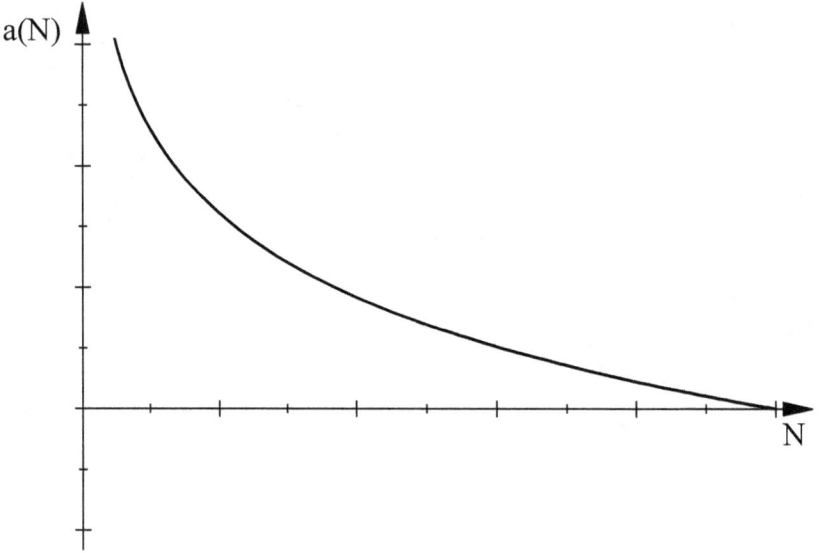

Figure 3.11: $a(N) = r \log \left(\frac{K}{N} \right)$

of fisheries. The model is given by

$$N^{n+1} = N^n + rN^n \left(1 - \frac{N^n}{K} \right) \triangle t. \tag{3.6}$$

Here

- $n =$ generation number,

- $N^n =$ population of nth generation,

- $\triangle t =$ time between generations,

- $K =$ carrying capacity.

In many ways there are parallels between discrete and continuous models since the discrete case can be rewritten as

$$\frac{N^{n+1} - N^n}{\triangle t} = rN^n \left(1 - \frac{N^n}{K} \right).$$

As $\triangle t \to 0$ the continuous time model is recovered. However, for fixed $\triangle t$, there are also some important differences, which we now look at carefully.

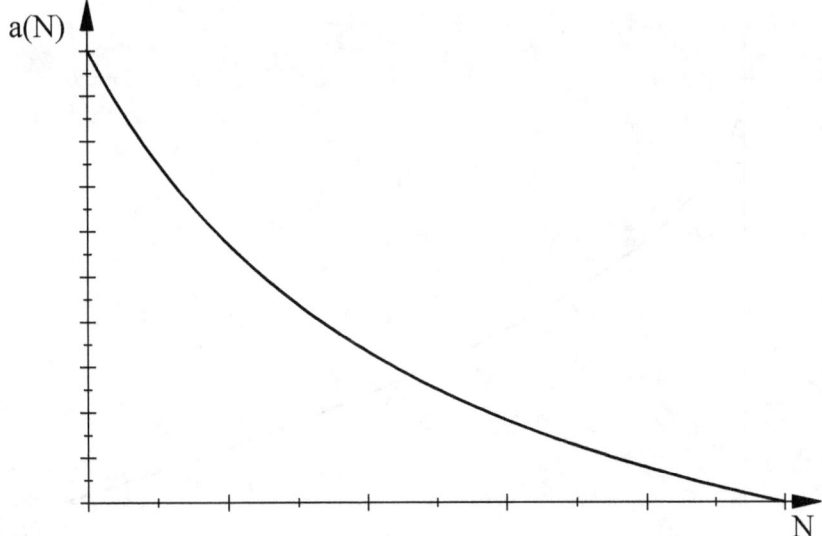

Figure 3.12: $a(N) = \frac{r(K-N)}{K+\alpha x}$

Equilibria. The discrete logistic model (3.6) is at equilibrium if $N^{n+1} = N^n$. This means

$$rN^n \left(1 - \frac{N^n}{K}\right) = 0$$

$$\Leftrightarrow$$

$$N^n \equiv 0 \quad \text{or} \quad N^n \equiv K \quad \text{for all } n.$$

The difference between the smooth approach to K in the logistic model and the behavior near K in the above discrete model is that the latter exhibits overshoot and undershoot oscillations.

A heuristic analysis. If $N^n < K$ then $N^{n+1} > N^n$ since

$$N^{n+1} = N^n + rN^n \left(1 - \frac{N^n}{K}\right) \triangle t$$

$$\text{and}$$

$$rN^n \left(1 - \frac{N^n}{K}\right) \triangle t > 0 \Leftrightarrow N^n < K.$$

Now suppose N^n is near K. Here "near" means $K - N^n$ is small. Isolate a small variable u^n by defining

$$u^n \text{ defined through } N^n = K + u^n \text{ so}$$
$$u^n := N^n - K.$$

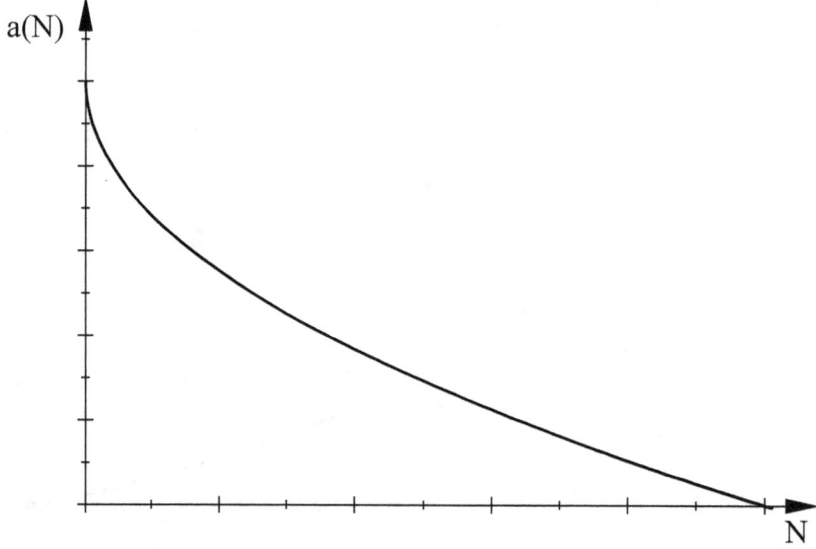

Figure 3.13: $a(N) = r\left(1 - \left(\frac{N}{K}\right)^{\theta}\right)$, here $\theta = 1/2$

With this definition, note in particular that

$$
\begin{aligned}
u^n &> 0 \text{ means overshoot: } N^n > K \\
u^n &< 0 \text{ means undershoot: } N^n < K.
\end{aligned}
$$

Replacing N everywhere in the discrete logistic equation by $N = K + u$ we have

$$
\begin{aligned}
u^{n+1} &= u^n - \frac{r\triangle t}{K} u^n (u^n + K) \\
&= u^n - r\triangle t u^n - \frac{r\triangle t}{K}(u^n)^2.
\end{aligned}
$$

Since u^n is small $(u^n)^2$ is negligible. Thus, near K,

$$
u^{n+1} \simeq u^n - r\triangle t u^n \simeq (1 - r\triangle t)u^n.
$$

Now if

$$
0 < 1 - r\triangle t < 1
$$

we have u^{n+1}, u^n have the same sign. On the other hand,

$$
\text{if } r\triangle t > 1 \text{ so } 1 - r\triangle t < 0
$$

then each step of

$$
u^{n+1} \simeq (1 - r\triangle t)u^n
$$

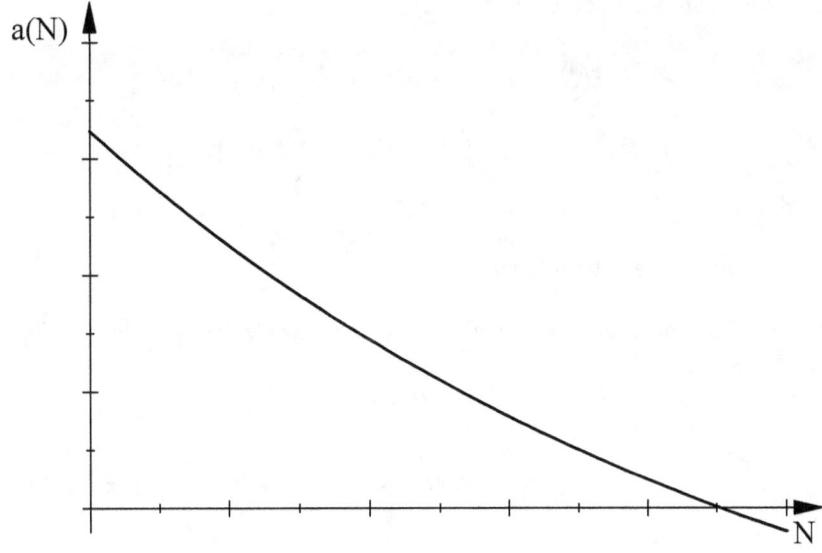

Figure 3.14: $a(N) = re^{1-N/K} - d$

changes sign. The population thus oscillates about K, first overshooting K then undershooting K then overshooting again and so on.

These oscillations around a mean carrying capacity are observed in populations of the total number of sheep in Tasmania[9], Davidson 1938, as well as in population levels in carefully controlled laboratory experiments.

Stability of the equilibrium $N = K$**.** We analyze stability of the equilibrium $N = K$ by applying a general theorem from numerical analysis (a useful version of the contraction mapping theorem).

Theorem 27 (Contraction Mapping Theorem) *Let f be continuously differentiable and K satisfy $K = f(K)$. Consider the iteration*

$$N^{n+1} = f(N^n).$$

If

$$|f'(K)| < 1$$

then for N^0 sufficiently close to K,

$$N^n \to K \ as \ n \to \infty.$$

---

[9] Try an internet search for a plot of this data.

To apply this theorem we calculate:

$$
\begin{aligned}
f(N) &= N + rN\left(1 - \frac{N}{K}\right)\triangle t \\
f'(N)|_{N=K} &= \left(1 + r\triangle t - \frac{2r\triangle t}{K}N\right)|_{N=K} \\
&= 1 - r\triangle t.
\end{aligned}
$$

Thus, we have proven the following.

Theorem 28 *The equilibrium state $N = K$ is stable for the discrete logistic model provided*
$$
|1 - r\triangle t| < 1,
$$
equivalently,
$$
0 < r\triangle t < 2.
$$

3.8   Overshoot and collapse models

"For the first time in history, we face the risk of a global decline. But we also are the first to enjoy the opportunity of learning quickly from developments in societies anywhere else in the world today, and from what has unfolded in societies at any time in the past."
   - Jared Diamond, Collapse: How Societies Choose to Fail or Succeed (2005).
   " ... the Rapanui experienced a tremendous upheaval in their social system brought about by a change in their island's ecology... By the time of European arrival in 1722, the island's population had dropped to 2,000–3,000 from a high of approximately 15,000 just a century earlier."
   - Barbara A. West, p. 684 in: Encyclopedia of the Peoples of Asia and Oceania. 2008, Infobase Publishing.
   The carry capacity of Rapa Nui[10] is quite different now as an arid land than when it was covered by forests.
   The logistic model is based on the assumption that the carrying capacity K is constant. When K is linked to a non-renewable resource this assumption is clearly false: increasing population increases the resource's depletion rate. Overshoot and collapse models correct the assumption of constant K by coupling the population level to an equation for resource depletion (or equivalently carrying capacity). This leads to models resembling

$$
N'(t) = r\left(1 - \frac{N(t)}{K(t)}\right)N(t),
$$
and
$$
K'(t) = g(N(t), K(t)).
$$

---

[10] This photo was taken by Hugo Krispyn. I appreciate his permission to use it here.

Figure 3.15: Rapa Nui was once covered by forests

These models are of both current and historical interest.

There have been a number of populations, societies and economies that grew, flourished, and then collapsed:

- **Rapa Nui.** Rapa Nui or Easter Island is believed to have been settled as early as 300 AD and there is reliable evidence of settlement by 900AD. Its population grew to a peak estimated to be as high as 30,000 people (at 450 people per square mile). There was a steep population crash in the 1600's to about 6000 people and by 1872 there were only 111 islanders living there (and living in desperate poverty). The collapse has been described in various ways but is usually linked to deforestation associated with boat building and for rollers to transport the great stone heads. Deforestation also led to severe drying out of the soil.

- **Pitcairn** and the **Henderson Islands**: These islands underwent similar population trajectories.

- The **Anasazi**. The Anasazi flourished around 600AD and developed high culture, arts, and complex agriculture based on irrigation. Total population levels are hard to estimate but they were geographically widespread and one site alone has been estimated at over 5000 people. By 1200 AD, possibly due to soil exhaustion and deforestation, they had completely disappeared.

- The **Maya** Collapse. At its peak, the Mayan population of Central America was estimated to be about 5,000,000 people with densities estimated to range from 250 to 1500 people per square mile. From 250 AD onwards,

Maya population increased exponentially to a peak in the 8th century followed by a total collapse by 909AD.

- The **Vikings in Greenland**. The Vikings in Greenland prospered and collapsed, disappearing completely.

These examples lead to the question:

What causes a flourishing society, economy and population to collapse suddenly?

This question is especially interesting since there is no way to know if we are all on the brink of a similar collapse. Naturally, just asking the question in this way presupposes that there is one unifying reason behind collapses of disparate societies. It is also just as possible that societies collapse for many different reasons and the real question is

Why does a society undergo explosive growth?

Nevertheless, many theories have been advanced[11] for overshoot and collapse events. There are currently two[12] with strong support:

1. Tainter's Complexity theory: As populations grow and become denser, societies become more complex. At some critical level of complexity, the whole economic and social structure becomes sensitive to external factors. Collapse is then inevitable when the right external stress occurs.

2. Diamond's Ecological collapse theory: A natural resource that supports a society's growth is overexploited for technology. Eventually the resource is depleted and the carrying capacity changes abruptly causing collapse.

There are several more concerning Rapa Nui that have less support.

3. Thor Heyerdahl's theory that it was settled by native South Americans who displaced less advanced peoples. This theory resulted in a wonderful adventure documented in the book *Kon Tiki*. However, it does not agree with genetic data. Its fit of other data is widely believed to depend on selection of which data to believe and reject. It is no longer considered valid.

4. The Atlantis theory is that Rapa Nui is the remnant of an advanced continent of Atlantis that sunk beneath the ocean. This theory has been successful only in selling books.

5. The extraterrestrial theory of Erich von Daniken. This theory has also been successful only in selling books.

We develop models of ecological collapse of the second type by coupling the level of some resource to a population's growth rate. These models may give a reasonable description of cases like Easter island. In complex economies they have been less accurate because they omit the economic incentive to develop substitutes for a scarce resource. We will develop three models of increasing complexity.

---

[11] It is nearly impossible, for example, to come up with a new and plausible theory about the collapse of the Roman empire.

[12] The oldest theory (not considered here) that once had wide acceptance is that collapse of republics is due to a decline in civic virtue.

3.8.1 Overshoot and Collapse for Non-Renewable Resources

" ... there are 12 so-called "mystery islands" in Polynesia. These islands were once settled by Polynesians but were unoccupied at the time of the European discovery. "
- J.A. Brander and M.S. Taylor 1998.
overshoot and collapse models use the following variables:

- $N(t) :=$ *Population level at time t,*

- $R(t) :=$ *Resource level at time t,*

- $a(R) :=$ *Population growth rate parameter,*

- $b :=$ *per capita rate of resource depletion.*

With these variables, the three assumptions of the first model are:
Assumption A1. logistic *growth occurs where the* carrying capacity $K = K(R)$ *depends on the resource level R.*
Assumption A2. *There is a small residual, subsistence, population with level denoted k_0. The carrying capacity grows linearly with R from that level onwards.*
Assumption A3. *The rate at which the resource is depleted is proportional to the population level.*
These assumptions lead, respectively, to the equations.

$$N'(t) = r \left(1 - \frac{N(t)}{K(R(t))} \right) N(t),$$

$$K(R) = k_0 + cR,$$
$$\text{and}$$
$$R'(t) = -bN(t).$$

Thus, the model is

$$N'(t) = r \left(1 - \frac{N(t)}{k_0 + cR(t)} \right) N(t),$$
$$\text{and}$$
$$R'(t) = -bN(t).$$

Analysis of the model. To plot the phase plane we begin by plotting the nullclines (in Figure 3.16 next). These are the two lines

$$N = 0 \quad \text{and} \quad N = k_0 + cR.$$

A few typical trajectories are plotted in Figure 3.17 below. The only long term

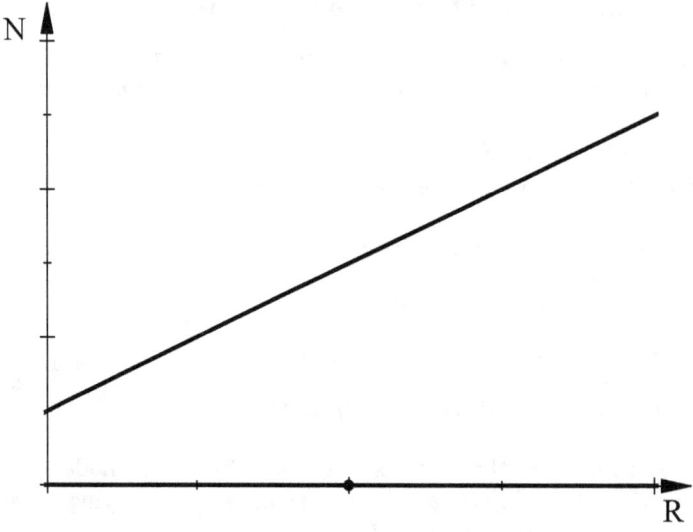

Figure 3.16:
Nullclines of the overshoot and collapse model
$$N' = r(1 - \frac{N}{k_0 + cR})N \text{ and } R' = -bN$$
$N =$ population level, $R =$ resource level.

outcome predicted is $R(t) \to 0$, and $N(t)$ converges to the residual (subsistence) population level k_0 as $t \to \infty$:

If the system starts from a small population and large resource level, the population will "explode" (grow rapidly) as the resource is depleted to a certain point thereafter the population will collapse to a small residual level and the resource will collapse to zero.

3.8.2   The Case of Renewable Resources

"Applying formal economic analysis to an archaeological mystery is an unusual activity for an economist."

  - J.A. Brander and M.S. Taylor, 1998.

The case of renewable resources is more complex. A basic model for this case was developed by Brander and Taylor [1998] based on standard (non-controversial) principles of both demographics and economics. Their model starts with the variables

- $N(t) :=$ population level,

- $R(t) :=$ resource level.

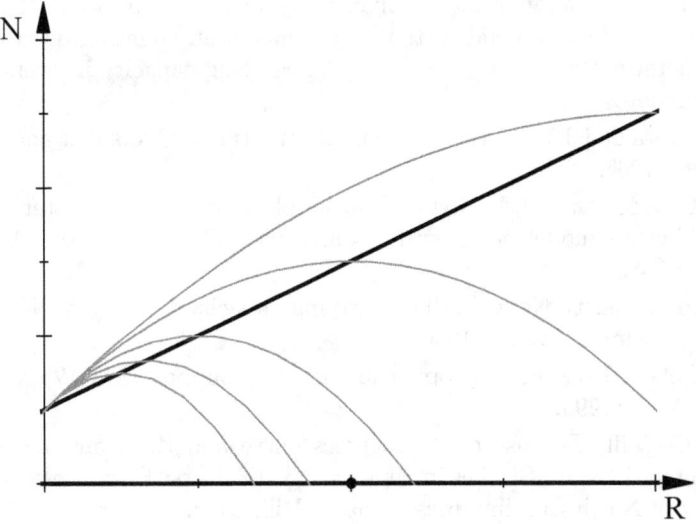

Figure 3.17:

Trajectories of the overshoot and collapse model
$$N' = r(1 - \frac{N}{k_0 + cR})N \text{ and } R' = -bN$$
$N(t) \to$ residual population level, $R(t) \to 0$.

Their model consists of the following equations

$$N'(t) = (b - d)N(t) + \alpha R(t)N(t),$$
$$\text{and}$$
$$R'(t) = rR(t)\left(1 - \frac{R(t)}{K}\right) - \beta N(t)R(t).$$

Here b, d are birth and death rates (with $b < d$), r is the resource renewal rate and K is its carrying capacity. The α, β are model coupling parameters. If $N = 0$ then the resource obeys the logistic equation while if $R = 0$ the population obeys a Malthusian population model. The response of this combination is more complex and depends on the relative parameter values.

3.9 References for Chapter 3

The most important conclusion in this chapter is a simple one. Exponential growth cannot continue indefinitely. It is resolved by some mechanism between complete collapse and gradual slowing to equilibrium. We have seen that simple models can give accurate predictions of aggregate population growth. Simple models (of complex phenomena) have thee great advantage of clarity and isolating the key effects that determine model behavior. One process in modeling in applied math was seen in the chapter: a simple model is often based on an assumption that some parameter (that really varies) is constant. The modeling process can then append to the simply model an equation describing how the parameter evolves, as was done for the carrying capacity in overshoot and collapse models.

L. Beierora and J.W. Forrester, Generic structures: Overshoot and Collapse, MIT, press, 1997.

J.A. Brander and M.S. Taylor, The simple economics of Easter Island: A Ricardo-Malthus model of renewal resource use, The Am. Econ. Review 88 (1998) 119-138.

carrying capacity Network, The carrying capacity briefing book, Washington, D C., volumes 1 and 2, 1996.

J.E. Cohen, How many people can the earth support?, W.W. Norton and Co., New York, 1995.

D.O. Cogwill, The use of the logistics curve and the transition model in developing nations, in: Studies in demography, (Editors: Boes, Desai and Hain) University of North Carolina press, Chapel Hill, 1970.

U. D'Ancona, The struggle for existence, Brill, Leiden, 1954.

J. Davidson, "On the Growth of the Sheep Population in Tasmania," Trans. R. Soc. S. Australia 62(1938): 342–346.

M.L. Deaton and J.J. Winebrake, Dynamic modeling of environmental systems, Springer, NY, 2000.

J. Diamond, Collapse: How societies choose to fail or succeed, Viking, NY, 2005.

P.R. Ehrlich, The population bomb, Valentine books, New York, 1968.

M.K. Hubbert, Energy from fossil fuels, Science 109(1945)103-109.

C.J. Krebs, Ecology: The Experimental Analysis of Distribution and Abundance, Harper and Row, New York, 1972.

T.R. Malthus, An essay on the principle of population, and a summary view of the principle of population, collected and edited by A. Few, Penguin, Baltimore, 1970.

J.R. Miner, Pierre-Francois Verhulst, the discoverer of the logistic curve, Human Biology, 5 (1933), 673-685.

R. Pearl and L. Reed, On the rate of growth of the United States population since 1790 and its mathematical representation, Proceedings National Academy of Science, 6 (1920) 275-288.

E.C. Pielou, An Introduction to Mathematical Ecology, Wiley, New York, 1969.

H. Pollard, Mathematical models of the growth of human populations, Cambridge University Press, Cambridge 1973.

V. Steffler, Long term forecasting and the problems of large scale wars, Futures 6(1974) 302-308.

J. Tainter, The collapse of complex societies, Cambridge U. Press, Cambridge, 1988.

3.10 Exercises for Chapter 3

1. If the resource is renewable the collapse models developed for non-renewable resources in the last section are inaccurate. Analyze the Brander and Taylor model for renewable resources:

$$N' = (b-d)N + \alpha RN,$$
$$\text{and}$$
$$R' = rR(1 - R/K) - \beta NR.$$

2. For the exact solution of the logistic equation

$$N(t) = \frac{K}{1 + \exp(-[rt + C])}.$$

Find C in terms of the initial population, r and K.

3. The fictional town of Cutbait, Wyoming has census data below. Estimate the growth rate constant in the Malthusian model using differences and averages. Calculate the exact value of the growth rate constant and compare the estimate to the exact value. Calculate the absolute, relative and percent relative error in the rate constant.

year	population
1997	12345
2012	23456

4. In some developing countries the birth rate seems to be proportional to N^k for some small positive number k. Solve the population growth equation rising from this birth rate by separating variables and interpret the results.

5. Find census data for a western European country and determine the parameters of the logistic model for that country. Use the years 1950, 1970, and 1990. Determine the validity of the model by looking up actual population data from 1900 (if available), 1940, 1960, 1980, 2000, and today and calculating the model's relative error.

6. The 2007 population growth rates of several countries are given below as the percentage change in population per year. Find the doubling or half life times (at least one of each) of four of them consistent with an

exponential population model.

Country	Growth Rate
Afghanistan	2.625
Bulgaria	-0.837
Canada	0.869
E.U.	0.160
Germany	-0.033
Slovenia	0.065
U.S.	0.894

percent change in population per year

7. Using census data from the United States in 1950, 1960, 1970, 1980, and 1990, and assuming a logistic model, find the parameter K using the graphical approach. What do you predict is the carrying capacity of the U.S.?

8. Verify that if $x(t) = 1/N(t)$ and $N(t)$ satisfies the logistic equation then

$$xN' + x'N = 0,$$

and $x(t)$ satisfies a linear, constant coefficient, ordinary differential equation. Solve this ordinary differential equation for $x(t)$.

9. Sketch the response of the logistic model of a single population when the carrying capacity is suddenly decreased. Begin with the logistic model with $N(0) > K$.

10. Consider the population of France, given below.

    (a) What are the units (hundreds of people, thousands, ten thousands ...) of the given population? Take averages and differences and develop a table of $N, N', N'/N$.

    (b) Plot carefully (on graph paper if by hand) the data on axes N against N'/N. Looking at the plot, give a range of predictions of the carrying capacity of France. Within this range give a most probable outcome and explain how you arrived at the estimate.

Year	:	'50	'60	'70	'80	'90
Population	:	41829	45670	50787	52811	56184

11. A simple generalization of the logistic model is

$$N' = aN + bN^2 + cN^3.$$

Using the US population from 1790, 1800, 1810 and 1820, find a fit to a, b, c and compare the new model's predictions to subsequent US populations and the logistic predictions.

12. Consider the population data below. (a) Plot the data on axes P'/P against P. Is the data logistic? Explain! (b) What is the population limit as $t \to \infty$? Explain!

$$\left|\begin{array}{cccccc} P(t) & : & 5 & 10 & 15 & 20 \\ P'(t) & : & 62.5 & 100 & 112.5 & 100 \end{array}\right|$$

13. The spread of a technological innovation is assumed to be proportional to the number of contacts between those using it and those not using it. If $x(t) = \#$ using the innovation and $N =$ total population size, assumed constant, find an equation for $x(t)$. Solve your model, sketch the solution and describe the solution behavior.

14. Rumors can spread in similar ways to technological innovations. Develop a model for the spread of a rumor. Explain its assumptions and analyze its predictions.

15. Reconsider Richardson's model of arms races. It supposes that each side can spend an unlimited amount on arms. Suppose instead that each block's economy has economic limitations with logistic type responses. Propose a modification. Sketch its phase portrait in one interesting case and compare its predictions to Richardson's model.

16. Cohen, in the appendix of his book, *How many people can the earth support?*, considers the population model with evolving carrying capacity

$$N' = rN(K - N),$$
$$K' = cN' = crN(K - N)$$
$$\text{where } c < 1.$$

(a) Show that $c > 1$ gives super exponential growth, which is experienced in some places.

(b) Pick some interesting parameters and sketch the phase portrait for $c < 1$.

(c) Cohen also describes the additional condition $c = c(t) = L/N(t)$ as the *J.S. Mill modification*. Repeat including this.

17. Trigger fish, commonly seen on reefs by skin divers, spend the first half of their life as females and then change to males for the second half of their life. Develop and analyze model for the population of trigger fish. Use as many components (variables) as necessary. Minimally this should require two variables ($F(t)$ and $M(t)$).

18. Consider the logistic equation for the population $N(t)$

$$N' = r\left(1 - \frac{N}{K}\right)N$$

Suppose that the carrying capacity $K = K(t)$ also depends on time and grows according to

$$K' = (b - K)K$$

where a, b are positive constants.

a. Sketch its phase portrait in the $N - K$ plane [for $N \geq 0, K \geq 0$]. Label the nullclines, trajectory directions, equilibria.

b. The system has an equilibrium at (b, b) : find the linearization there and classify it.

19. The population of Allegheny County in Pennsylvania, below, has been decreasing according to the census data:

Year	Population (in millions)
1970	1.605
1990	1.336
1994	1.321
1995	1.310

(a) Fill in systematic approximations for the blank entries.

Year:	1970	1980	1990	1995	2000
N	1,605,229	_____	1,336,449	_____	1,281,666
N'	*	_____	*	_____	*
N'/N	*	_____	*	_____	*

(b) Using this table, predict the final population of Allegheny county ($lim_{t \to \infty} N(t)$) with the logistic model.

20. In the 1970's and 1980's there was a tremendous investment in codes to replace wind tunnel tests of airplane designs by numerical simulations, called "*the numerical wind tunnel*". For several years there was great progress which tapered off, see the data below.

year	1978	1983	1990	1995	1999
#tests	77	38	18	11	10

year & # wind tunnel tests

Model this decline with a logistic model with $N(0) > K$. Predict the ultimate number of tests that will be needed (if current trends continue). It is interesting to note that the actual number of tests has been stuck at 10 from 1999 to 2019.

21. The initial value problem $x' = ax - bx^2$, $x(0) = x_0$, where $a > 0, b > 0, x_0 > 0$ describes a logistic growth process. What is the limit $lim_{t \to \infty} x(t)$?

Chapter 4

Errors in Models

"As far as the laws of mathematics refer to reality, they are not certain; and as far as they are certain, they do not refer to reality."
— *Albert Einstein*

Descriptive models are mathematical toys used to do thought experiments. Beyond thought experiments, prediction is the essential aim of useful modeling. Prediction means the *error in the prediction* becomes of essential interest. As error is an essential feature, there are trademarked terms.

Definition 29 *The error, relative error and % error are, respectively,*

$$
\begin{aligned}
error &= \textit{True Value - Approximate Value} \\
relative\ error &= \left| \frac{\textit{True - Approximate}}{\textit{True}} \right| \\
\%\ error &= \textit{Relative Error} \times 100.
\end{aligned}
$$

Loosely speaking, the view is:

- In interesting cases, the error is *essential but unknowable*[1],

- The relative error is the correct error measure to study and report[2].

For our purposes herein, a dynamic model consists of a collection of difference or differential equations with parameters, data used to determine the parameters, data used to provide initial conditions and a numerical method to generate predictions from the collection of equations. At every step errors are introduced. Collections of differential equations usually err by extreme oversimplification of complex phenomena Thus all model predictions are contaminated by the modeling process. It is useful to keep in mind the Hamming epigram:

---

[1] Unknowable because when the aim is prediction and the problem is beyond a toy problem, the true values are not known.

[2] The relative error is correct for two reasons. First, what error level is good depends on problem size; the relative error scales out the problem size automatically. Second the relative error tells how many significant digits of the prediction are correct: relatve error $= 10^{-k}$ means that k significant digits are trustworthy.

"All models are wrong. Some models are useful."
-R. Hamming

Data used to determine model parameters contain, like all data, errors. Thus model parameters used are incorrect and their errors cascade through all model predictions. Numerical methods induce their own collection of errors. Fortunately, numerical analysis is one of the few areas of mathematics where errors have been systematically studied and controlled. Modern numerical methods, based on self-adaptive, variable step and variable order methods, are quite reliable. Thus normally numerical errors are far smaller than other sources[3] and other sources should thus be studied using reliable numerical methods. Even with all other sources controlled, the error in model predictions at latter times that arise from the growth of small errors in initial conditions can often invalidate any model predictions. In this chapter we give a very short presentation of four topics related to sources of errors in models and their treatment.

Calibration: How to use linear least squares / linear regression to determine parameters.

Conditioning: *The magnification of errors made by using approximate input values in otherwise exact formulas.*

Sensitivity: *errors in predictions due to parameters are studied through sensitivity equations, Section 1.*

Predictability: *errors in model predictions arising from initial condition errors are quantified by studying the predictability horizon through ensemble simulations, Section 2.*

4.1   Linear least squares / linear regression

To be concrete we start with the logistic model

$$N' = r\left(1 - \frac{1}{K}N\right)N$$

The values of r and K must be specified before predictions can be made with the model. Since the model has a closed-form, exact solution, the values of r and K can be determined using the exact solution and exact data. If, on the other hand, we are developing a model without a closed-form, exact solution we would have to resort to some other technique. The most common and most basic approach is to use a *linear least squares fit*, invented by C.F. Gauss and also called a *linear regression*. Since most scientific calculators will do this reliably we will explain the idea without extensive hand-calculated examples.

The first step is to identify a possible linear relationship in which the sought parameters are related to the slope and intercept. For the logistic equation this

---

[3]Spectacular errors do continue to occur by choosing the wrong numercal methods for the problem under consideration and then trusting its results. Students who want to see the fruits of mathematics in the world are encouraged to study numerical analysis.

is

$$\frac{N'}{N} = -\frac{r}{K}N + r \quad \Leftrightarrow \quad y = mx + b$$
$$y = \frac{N'}{N} \quad , \quad x = N$$
$$m = \text{slope} = -\frac{r}{K} \quad , \quad b = \text{intercept} = +r.$$

Thus data in the form $(N, \frac{N'}{N})$ is a collection of $x - y$ points

$$(x_1, y_1), \cdots, (x_n, y_n)$$

in which noise and error are expected. If the data points are plotted we expect it should approximate a line (if the model is accurate). and thus seek the line of best fit minimizing the deviations in some sense. Finding this best fit line is done by a linear least squares fit, a technique invented by C.F. Gauss to predict the paths of celestial bodies (that he referred to as "clods of earth").

Definition. *The residual, a measure of how close the point (x_i, y_i) is to the line $y = mx + b$, is*

$$r_i = mx_i + b - y_i.$$

A linear least squares fit is done by picking the line (i.e., picking m and b) that minimizes the sums of the squares of the residuals of all points. Define

$$f(m, b) := \sum_{i=1}^{n} (mx_i + b - y_i)^2.$$

Here x_i, y_i are numbers and the variables are m, b. The minimizer is where the partial derivatives vanish

$$\frac{\partial f}{\partial m}(m, b) = 0 \quad \text{and} \quad \frac{\partial f}{\partial b}(m, b) = 0.$$

Since the function to be minimized is quadratic, its derivatives are linear. This is a 2×2 linear system. Calculating we get the linear system

$$\begin{bmatrix} \sum_{i=1}^{n} x_i \cdot x_i & \sum_{i=1}^{n} x_i \cdot 1 \\ \sum_{i=1}^{n} x_i \cdot 1 & \sum_{i=1}^{n} 1 \cdot 1 \end{bmatrix} \begin{bmatrix} m \\ b \end{bmatrix} = \begin{bmatrix} \sum_{i=1}^{n} x_i \cdot y_i \\ \sum_{i=1}^{n} 1 \cdot y_i \end{bmatrix}.$$

It can be shown that this 2×2 matrix is always invertible if there are at least two distinct points. Thus linear least squares fit can always be calculated. It does not mean it is always an accurate representation of the data. This is usually assessed two ways:

1. plotting the data and the line and looking at them, and

2. calculating the Gauss criteria for goodness of fit for which smaller values mean a better fit:

$$S^2 := \frac{\sum_{i=1}^{n} (mx_i + b - y_i)^2}{n - 2}.$$

While begun by a great mathematician, the whole area of finding accurate fit to data was thereafter highly developed by statisticians (of various types). Accordingly, the place to learn more is in (various) regression courses on in (one of the various types of) statistics departments.

4.2   Conditioning

"Although this may seem a paradox, all exact science is dominated by the idea of approximation."
  -Bertrand Russell (1872-1970)

Errors occur when using exact data with a simplified model as well as when using an exact model with uncertain data. An example of this occurs when we used population data from 1790 and 1800 to calculate the growth rate constant for the Malthusian model of the US population. Since the 1800 population has error itself, it is good to ask how accurate an exact formula is when using approximate values. The answer of course depends on the formula and the values used. Fortunately, the exact dependence is calculable as follows.

Restating (simplifying the problem by making it more general), the problem is:

> *For a function $y = f(x)$, what is the relative error in y (the output) computed from an approximate value of x (the input) with a given relative error?*

This question is answered by the **condition of a function at a point**.

Definition 30 *Let $y = f(x)$. Suppose $f(x_0) \neq 0$. The condition of $f(x)$ at $x = x_0$ is **the relative error in $f(x)$ divided by the relative error in x** :*

$$\mathit{condf}(x_0) := \lim_{x \to x_0} \frac{\left| \frac{f(x_0) - f(x)}{f(x_0)} \right|}{\left| \frac{x_0 - x}{x_0} \right|}$$

Example 31 *For example, if a is known to 5 significant digits (relative error 10^{-5}) and $\mathit{condf}(a) = 1000$, then*

$$\left| \frac{f(a) - f(x)}{f(a)} \right| \simeq 1000 \times 10^{-5} = 10^{-2}$$

which means $f(a)$ is only known to 2 significant digits.

The value of $\mathit{condf}(x_0)$ is calculated as follows.

Lemma 32 *We have*

$$\mathit{condf}(a) = \left| \frac{a f'(a)}{f(a)} \right| .$$

Proof. We have

$$\frac{\left| \frac{f(x_0) - f(x)}{f(x_0)} \right|}{\left| \frac{x_0 - x}{x_0} \right|} = \left| \frac{f(x_0) - f(x)}{x_0 - x} \right| \times \left| \frac{x_0}{f(x_0)} \right| \to \left| \frac{x_0 f'(x_0)}{f(x_0)} \right| .$$

∎

We can now apply this to the formula for the exact growth rate constant

$$a(N_1) = \frac{1}{t_1} \ln \left(\frac{N_1}{N_0} \right).$$

We calculate

$$\begin{aligned}
cond\ a(N) &= \left| \frac{Na'(N)}{a(N)} \right| = \cdots \\
&= \left| \frac{1}{\ln \left(\frac{N}{N_0} \right)} \right|.
\end{aligned}$$

Observe that *cond a(N) → ∞ as N → N₀. For the Malthus model, the closer the data points, the more the inevitable error in the data used to calibrate the model is magnified in the model's solution.*

For 1790 and 1800 we have

$$cond\ a(5308483) = \left| \frac{1}{\ln \left(\frac{5308483}{3929214} \right)} \right| = 3.3.$$

Thus, the error in a using $N(1790)$ and $N(1800)$ is expected to be 3.3× the error in $N(1800)$.

4.3   Model Sensitivity

"The road to wisdom? - Well, it's plain
And simple to express:
Err
and err
and err again,
But less
and less
and less."
- Piet Hein

We consider only one specific model: the logistic equation

$$N' = r \left(1 - \frac{N}{K} \right) N \text{ for } t > 0,$$
$$N(0) = N_0$$

To calculate a solution to the model the initial condition N_0 and the 2 parameters r (the initial growth rate) K (the carrying capacity). Since r, K are derived from population data, we consider the model solution to be a function of r, K and t:

$$N = N(t; r, K).$$

The parameters r, K are determined by fitting the model to data and thus the values used in any prediction are uncertain. This uncertainty could be magnified or damped by the formula used to calculate the parameters. Sensitivity equations quantify the effect of a small error in a model parameter upon the model's solution.

Definition 33 *The sensitivities of the prediction of the logistic model with respect to r and K are, respectively,*

$$
\begin{aligned}
S_r(t) \quad &: \quad = \frac{\partial N}{\partial r}, \\
S_K(t) \quad &: \quad = \frac{\partial N}{\partial K}.
\end{aligned}
$$

More generally, given an ODE with a parameter α

$$ y' = f(t, y; \alpha), $$

the sensitivity of the solution $y(t)$ with respect to α is

$$ s(t) = \frac{\partial y}{\partial \alpha}. $$

A model sensitivity is valuable because it gives insight into how strongly the solution responds to changes in the parameter studied. This includes giving insight into how the model's solution responds to errors in data used to calibrate the model. The way $s(t)$ is evaluated is through implicit differentiation of the equation with respect to the parameter. For the general initial value problem we calculate the sensitivity equation to be the coupled system, nonlinear in y but linear in s, to be

$$
\begin{aligned}
y' &= f(t, y; \alpha), \\
s' &= f_y(t, y; \alpha)s + f_\alpha(t, y; \alpha).
\end{aligned}
$$

Explicit formulas for the solution of the sensitivity equation are seldom possible. Fortunately, as we have just seen, it is very easy to derive an explicit equation for sensitivities. These equations can be studied by the same techniques used to study the original equation.

   Example: Sensitivity of logistic equation. For the logistic equation we now work through the sensitivity equation. Taking $\frac{\partial}{\partial r}, \frac{\partial}{\partial K}$ of the logistic equation (i.e., differentiating implicitly) gives:

$$
\frac{\partial}{\partial r} \left[\frac{d}{dt} N = r \left(1 - \frac{N}{K} \right) N \right] \quad \Rightarrow
$$

$$
\begin{aligned}
\frac{d}{dt} \left(\frac{\partial N}{\partial r} \right) &= \frac{\partial}{\partial r} \left[rN - \frac{r}{K} N^2 \right] \\
&= 1N + r\frac{\partial N}{\partial r} - 1\frac{1}{K} N^2 - \frac{r}{K} 2N \frac{\partial N}{\partial r}
\end{aligned}
$$

Thus $S_r(t) := \frac{\partial N}{\partial r}$ satisfies

$$\frac{d}{dt}S_r = r(1 - \frac{2}{K}N)S_r + N(1 - \frac{N}{K}).$$

Notice that, once $N(t)$ is calculated this is a linear equation for the sensitivity with respect to r. It turns out that implicit differentiation always yields linear equations for sensitivities. This is because sensitivity is a theory explaining the effect of small errors. The sensitivity equation is solved as part of a coupled system. Usually the initial sensitivity is taken to be $S(0) = 0$. This gives

$$N' = r\left(1 - \frac{N}{K}\right)N \text{ with } N(0) = N_0,$$

$$S_r' = r(1 - \frac{2}{K}N)S_r + N(1 - \frac{N}{K}) \text{ with } S(0) = 0.$$

Modern numerical methods have no difficulty with such a system. The normal inference is that when and where sensitivities are large, parameter errors make model predictions questionable.

The sensitivity with respect to the carrying capacity is even more interesting and important. Its equation is found exactly the same way: by implicit differentiation, to be

$$S_K' = r(1 - \frac{2}{K}N)S_K + r\frac{N^2}{K}.$$

sensitivity equations can easily be appended to any numerical method to estimate the quantity of interest. For example, for the logistic equation the system

$$N' = r\left(1 - \frac{N}{K}\right)N \quad , \quad N(0) = N_0,$$

$$S_K' = r\left(1 - 2\frac{N}{K}\right)S_K + r\frac{N^2}{K^2} \quad , \quad S_K(0) = 0.$$

is autonomous. Its slope field and phase portrait can easily be constructed by many numerical methods. The slope field is depicted next in Figure 4.1 for $r = 1, K = 5$. A phase portrait of the system $(x, y) = (N, S)$ is presented in Figure below for the same values $r = 1, K = 5$. (This phase portrait was done with Polking's Pplane program.) The plot shows that the sensitivity is greatest when the population is $N(t) = K/2$. As $t \to \infty, x \to K$. Thus the sensitivity as $t \to \infty$ approaches $S \to \partial K/\partial K = 1$, as seen in the figure.

We conclude that before action is taken based on a prediction of any model, the sensitivity with respect to all model parameters should be evaluated by solving the model plus its sensitivity equations.

For the logistic model this is the system:

$$N' = r\left(1 - \frac{N}{K}\right)N \text{ with } N(0) = N_0,$$

$$S_r' = r(1 - \frac{2}{K}N)S_r + N(1 - \frac{N}{K}) \text{ with } S_r(0) = 0,$$

$$S_K' = r(1 - \frac{2}{K}N)S_K + r\frac{N^2}{K} \text{ with } S_K(0) = 0.$$

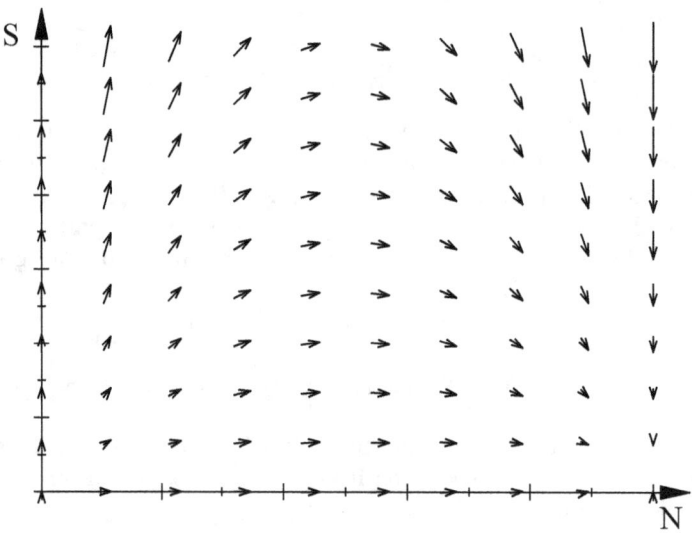

Figure 4.1: Slope field of population $N(t)$ & sensitivity $S(t)$

4.4 Predictability: an Introduction

"If we knew exactly the laws of nature and the situation of the universe at the initial moment, we could predict exactly the situation of the same universe at a succeeding moment. But even if it were the case that the natural laws had no longer any secret for us, we could still only know the initial situation approximately. If that enabled us to predict the succeeding situation with the same approximation, that is all we require, and we should say that the phenomenon had been predicted, that it is governed by laws. But it is not always so; it may happen that small differences in the initial conditions produce very great ones in the final phenomena. A small error in the former will produce an enormous error in the latter. Prediction becomes impossible, and we have the fortuitous phenomenon." - Henri Poincaré, 1903 "Science and Method"

Let us assume for this section all other sources of error are well controlled except for how inevitable but small errors in the initial conditions evolve in the model. Thus for the model

$$y' = f(t, y), t > 0,$$
$$y(0) = y_0,$$

predictability addresses the question:

What is the evolution of the error

$$|y(t) - y_{true}(t)|$$

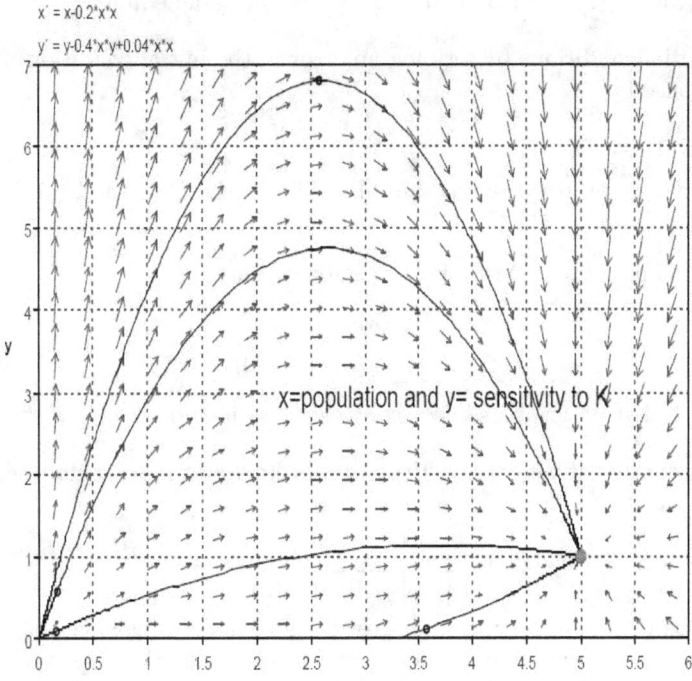

x´ = x-0.2*x*x

y´ = y-0.4*x*y+0.04*x*x

x=population and y= sensitivity to K

Figure 4.2:   Solution and Sensitivity wrt K
                for the choices $r = 1, K = 5$

in the model's predictions when there is a small error in the initial value y_0?

Lorenz detailed two predictability problems. The second is generally more important and less studied.

1. Predictability of the first kind: *Forecast, within an acceptable tolerance, the complete solution $y(t)$ to the system of differential equations at some future time, the problem above.*

2. Predictability of the second kind: *Forecast, within an acceptable tolerance, averages of the solution $y(t)$ to the system of differential equations at some future time.*

The description and understanding of predictability (and "chaos" which limits predictability) begins with rearranging what is known about exponential growth into statements about errors. To begin, consider simple exponential growth

$$y' = ay, \quad a > 0, \, y(0) = y_0.$$

This has fixed doubling time

$$T_{double} = \frac{\ln 2}{a}.$$

Alternately, let $y_1(t), y_2(t)$ be two solutions of the same equation with slightly different initial conditions (describing an error in the initial condition). Denote their difference by

$$v(t) := y_1(t) - y_2(t).$$

If one of $y_1(t), y_2(t)$ uses the true initial condition the difference, $y_1(t) - y_2(t)$, is the error. The error $v(t)$ satisfies (by subtraction)

$$
\begin{aligned}
v' &= av, \\
v(0) &= y_1(0) - y_2(0)
\end{aligned}
$$

The solution that determines how the error grows in time is $v(t) = e^{at}(y_1(0) - y_2(0))$. Alternately, if we *observe* the two solutions $y_1(t), y_2(t)$ then the growth rate is

$$a = \frac{1}{t} \ln \left| \frac{y_1(t) - y_2(t)}{y_1(0) - y_2(0)} \right|.$$

For solutions of simple, exponential growth models a, so calculated, is a constant. However, for any model and any two solutions $y_1(t), y_2(t)$ we can insert the values in the above formula (even without knowing the model if we have data on solution values) and calculate a, *which will now depend on t* so $a = a(t)$. Given the value of $a(t)$, we insert this into the above doubling time formula to get an observed, experimental value for a doubling time of initial errors.

Definition 34 *The observed doubling time of the initial perturbations $y_1(0), y_2(0)$ is*

$$T_{double}(t) := \frac{\ln 2}{\frac{1}{t} \ln \left| \frac{y_1(t) - y_2(t)}{y_1(0) - y_2(0)} \right|}.$$

This can be extended to any average or 'statistic' computed from $y(t)$. We observe that *larger doubling times means greater predictability* and *smaller doubling times means less predictability*.

Doubling times are an indirect measure of predictability. The $\varepsilon-$predictability horizon, T^ε, is a direct measure calculated from doubling times.

Proposition 35 *Let $\varepsilon > 0$. For the exponential model $y' = ay$, if we want*

$$|y_1(t) - y_2(t)| < \varepsilon$$

then we must have

$$t \leq T^\varepsilon := \frac{1}{a} \ln \left(\frac{\varepsilon}{|y_1(0) - y_2(0)|} \right).$$

For a more general model, $a = a(t)$ and we therefore define the ε−predictability horizon, replacing the number a by $a(t)$ calculated from model solutions as follows.

Definition 36 *Let a general model have solutions* $y_1(t), y_2(t)$ *from different*

initial conditions $y_1(0), y_2(0)$. *The* ε−*predictability horizon of the model,* $T^\varepsilon(t)$,

is then

$$
\begin{aligned}
T^\varepsilon(t) \quad : \quad &= \frac{1}{a(t)} \ln \left(\frac{\varepsilon}{|y_1(0) - y_2(0)|} \right) \quad \text{where} \\
a(t) \quad &= \quad \frac{1}{t} \ln \left| \frac{y_1(t) - y_2(t)}{y_1(0) - y_2(0)} \right|
\end{aligned}
$$

If $a(t) > 0$ then the system will have a finite ε−predictability horizon. It is an observation of Lorenz that T^ε depends logarithmically upon the initial error $y_1(0) - y_2(0)$. Thus, $a(t) >> 1$ (i.e., $a(t)$ is very large) it is practically impossible

to make valid predictions over longer time intervals simply by specifying the initial condition to greater accuracy. This is the first hint of the problem of predictability.

4.4.1   The problem of predictability

"A chaotic system is one which is exactly predictable but not approximately predictable." - E.N. Lorenz

The question of predicting $y(t)$ of the last section is fundamental but also has a fundamental error in how it is asked that we now correct. The error in phrasing is that it addresses the absolute error

$$
\text{absolute error} := |y(t) - y_{true}(t)|
$$

which is irrelevant since the size of the solution is unknown. It must be adjusted to address the relative error

$$
\text{relative error} \quad := \frac{|y(t) - y_{true}(t)|}{|y_{true}(t)|}.
$$

It is easy to check that the size of the relative error tells exactly how many significant digits of accuracy exist:

If the relative error is 10^{-k} *then the approximation has* k *accurate significant digits.*

The difference is apparent for the simple exponential growth problem $y' = ay$. For this problem, because the solution grows as fast as the separation between solutions, the relative error does not grow at all:

error grows exponentially   : $|y(t) - y_{true}(t)| = e^{at}|y(0) - y_{true}(0)|$

Relative error bounded   : $\dfrac{|y(t) - y_{true}(t)|}{|y_{true}(t)|} = \dfrac{e^{at}|y(0) - y_{true}(0)|}{e^{at}|y_{true}(0)|}$
$$= \frac{|y(0) - y_{true}(0)|}{|y_{true}(0)|}.$$

Thus linear problems are relatively predictable because the solution grows as fast as the error.

Obviously the real danger is for a nonlinear problem where solutions can separate exponentially but remain bounded. This is the classical definition of chaos given by Lorenz.

Definition 37 *A system is chaotic if*

1. Perturbations of the initial conditions grow exponentially until they differ in their first significant digit, and

2. All solutions $y(t)$ remain uniformly bounded in time: for some constant C, $0 < C < \infty$,

$$\sup_{0 \leq t < \infty} |y(t)| \leq C < \infty.$$

and if the solution's short time or transient behavior is ignored, the bound is independent of the initial condition. In detail, there is a fixed upper bound constant C such that, for any y_0, for the solution with $y(0) = y_0$, there is a T with

$$\sup_{T \leq t < \infty} |y(t)| \leq C < \infty.$$

The first condition is studied by solving the system for several initial conditions and plotting the data

$$a(t) \;\; = \;\; \frac{1}{t} \ln \left| \frac{y_1(t) - y_2(t)}{y_1(0) - y_2(0)} \right|,$$
$$T^{\varepsilon}(t) \;\; : \;\; = \frac{1}{a(t)} \ln \left(\frac{\varepsilon}{|y_1(0) - y_2(0)|} \right).$$

The second condition is usually built into the model and reflects some physical feature, such as the setting containing a total finite amount of something.

Model's of real systems can have hundreds of thousands to millions of variables. Thus simplified, toy, models are often used to try to understand predictability in a comprehensible, qualitative way. Typically these describe the error in a solution

$$E(t) \simeq |y(t) - y_{true}(t)|.$$

The first model used by Lorenz was the logistic equation. This was soon modified slightly to the following. Assuming the initial error and total, final error are estimated we expect $0 < E_{\min} < E(t) \leq E_{\max} < \infty$. An equation which describes how an initial error evolves to a maximum is

$$
\begin{aligned}
E'(t) &= s(E_{\max} - E(t))(E(t) - E_{\min}), \\
E(0) &= E_0 \ (> E_{\min}).
\end{aligned}
$$

The possibility of chaos means for modeling that before action is taken on the predictions of any model, an ensemble calculation should be performed (solving the model for various initial conditions) and the model predictability horizon calculated[4]. Model predictions occurring after the model predictability horizon should be regarded as equivalently good as if generated by a random number generator.

Chaos does not occur in an autonomous system of two equations. However, *chaos has turned out to be the typical and expected case for systems of three of more equations.* The importance of ensemble simulations to study predictability will increase for the foreseeable future.

4.5 References for Chapter 4

Where mathematics is needed a "model" is a black box where numbers go in and predictions come out. As such, the error in the predictions is the key metric for assessing value of a model. When a model gives predictions as time evolves the error in the model will often grow with time. In this case a model can start with accurate predictions but then have a finite predictability horizon. Beyond that horizon, the model acts as a mathematically attractive random number generator. We have developed several ways to come to calibrate models, to understand model error that arises from data error and to evaluate and extend the predictability horizon of a model. For deeper study of these issues, please see:

E. Kalnay, Atmospheric Modeling, Data Assimilation and Predictability, Cambridge Univ. Press, 2003.

LG Stanley and DL Stewart, Design sensitivity analysis: computational issues of sensitivity equation methods, SIAM, 2002.

4.6 Exercises for Chapter 4

1. Derive the sensitivity equation $S'_K = r(1 - \frac{2}{K}N)S_K + r\frac{N^2}{K}$.

2. For the logistic equation with $r = 1, K = 10$, plot the difference between solutions with slightly different initial conditions. Then calculate explicitly and plot $a(t)$ and $T^\varepsilon(t)$.

---

[4]Often the prediction of solution averages is needed. Then the predictability horizon of those averages should be computed. This can be much different that the predictability horizon of the solution.

3. Consider the logistic equation

$$y' = ay\left(1 - \frac{y}{K}\right), t > 0.$$

Let $y_{1,2}(t)$ be two solutions with different initial conditions and $u(t) = y_1(t) - y_2(t)$.

(a) Show that for large enough time

$$\frac{u'}{u} < 0$$

so that $y_{1,2}(t)$ squeeze together exponentially fast.

(b) Show that for small time $y_{1,2}(t)$ separate exponentially fast by showing

$$\frac{u'}{u} > 0 \text{ while } \frac{y_1(t) + y_2(t)}{2} < K.$$

(c) If someone analyzing this logistic population growth incorrectly supposes it is exponential, they would calculate the growth rate (and be surprised it is not constant) by:

$$a = a(t) = \frac{1}{t} \ln\left(\frac{y(t)}{y(0)}\right) \text{ for any } t > 0.$$

Sketch of the behavior of this function $a(t)$ against t for $y(t)$ a solution of the logistic equation.

(d) Repeat part c for the predictability window plotted against t and the epsilon predictability horizon plotted against t.

4. Consider the toy error equation

$$
\begin{aligned}
E'(t) &= s(E_{\max} - E(t))(E(t) - E_{\min}), \\
E(0) &= E_0 \ (> E_{\min}).
\end{aligned}
$$

Show that this can be rewritten in the form

$$E'(t) = (\alpha E(t) + \gamma)(1 - \frac{E(t)}{E_{\max}}).$$

Plot the solution for $\gamma = 0$ (the logistic equation) and for $\gamma > 0$.

5. Suppose all the details of the solution of $y' = f(t, y)$ are not needed but only some average denoted $A(y(t))$. Derive a formula for the ε−predictability horizon of those averages. Pick a specific time or other average and make your formula concrete for that average.

Chapter 5

The Lotka-Volterra Model

5.1   Population Oscillations

"To test a recently developed predator-prey model against reality, I chose the well-known Canadian hare-lynx system. ... The correlation between the model and the empirical data gives some idea about the general worth of the model....."
- M.E. Gilpin, 1972.

The development of models of interacting species was strongly motivated by the challenge of understanding some puzzling patterns in their population fluctuations.

Mystery 1: Reduced fish harvesting lowered population.

In 1926 Umberto D'Ancona[1] (an Italian biologist) approached Vito Volterra (a mathematician famous for his work on differential and integral equations including what we now call *Volterra equations*) about a pattern he noticed in fishery data. The percent of the total catch by Italian fishermen in the Adriatic of the predator species Salecians (comprising sharks and rays) was as follows.

1914	'15	'16	'17	'18	'19	'20	'21	'22	'23
11.9%	21.4	22.1	21.2	36.4	27.3	16.0	15.9	14.8	10.7

% of total catch of sharks and rays

The mystery was that, during WWI when the war inhibited fishing (harvesting of the prey species), there was a large and consistent relative *increase in predator* fish populations and *decrease in prey* fish populations. This lead to the questions:

How does harvesting affect the composition of fish populations?
and
Why are the effects different for predators and prey species?

---

[1] Volterra was his father in law. D'Ancona expected that not harvesting the food fish would increase their % of the harvest and thus decrease the % of incidentally harvested predator fish.

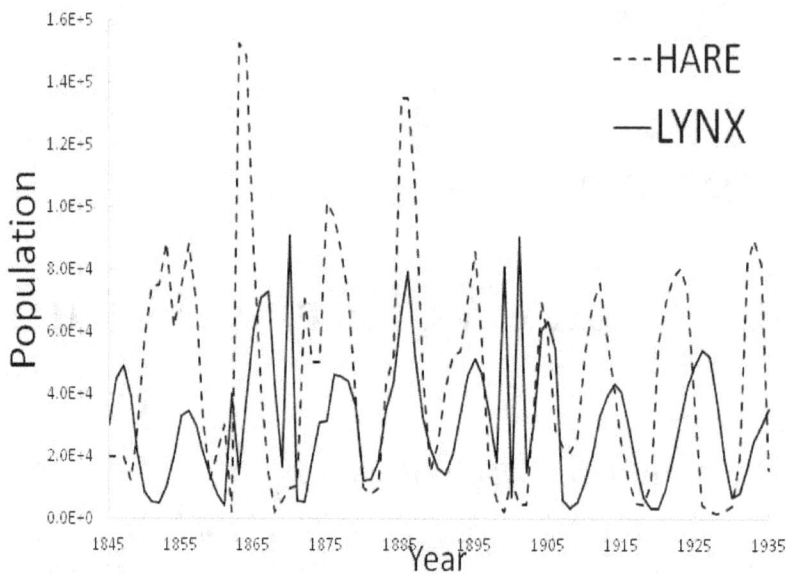

Figure 5.1:   Cycles appear in the Hare-Lynx pelt data
                The period is approximately 10 years.

Mystery 2: Spraying with DDT increased pest population.

The cottony cushion insect arrived in the US in 1868 from Australia and almost destroyed the U. S. Citrus industry. Its natural predator, the ladybird beetle, was then imported to control it. When DDT was introduced into farming practice, hoping to further reduce the population of cottony cushion insects, the overall effect was to *increase the population of the cottony cushion insect* and *decrease that of the lady bird beetle*. The questions are:

Why does this happen?

and

How should biological and chemical controls be combined?

Mystery 3: Lynx and snowshoe hare populations.

The Hudson Bay company has records of numbers of furs harvested by trappers dating back to even before 1840. These records show a persistent 10 year cycle in harvesting (and thus in population levels) that does not settle down to some sort of stable ratio but seemed to be built into the interactions of the two species. The cyclic behavior is evident in Hudson Bay Company's pelt data of snowshoe hare and lynx, plotted next in Figure 5.1. This led to the question:

Why do cycles with persistent periods occur in predator and prey population levels?

5.2 The Predator Prey Model

"To perform the test, the derivatives in equations (1a) and (1b) were approximated by the per year changes in lynx and hare densities. Next the parameters in these equations were adjusted so that the sum of the squares of the error... is minimized." -M.E. Gilpin 1974.

The predator prey model of two interacting populations was developed independently by Lotka and Volterra. The classic examples of predator-prey interactions are lynx and snowshoe hares and (in text books at least) foxes ($F(t)$) and rabbits ($R(t)$). In either case, the assumptions are as follows.

Assumption A1. *The prey have an infinite supply of food but the prey is the only food of the predator.*

Assumption A2. *In the absence of the other, the predator will undergo a pure death process and the prey a pure birth process.*

Assumption A3. *The number of kills of prey by predators is proportional to the number of contacts between predators and prey, which is proportional to the product of the two populations.*

The contact assumption is one key to the Lotka-Volterra model. Imagine putting two groups of blind, drunk rabbits and foxes in a room. If both stagger around at random there will be a certain number of contacts. Doubling either would double the number of contacts and doubling both increases the number 4 fold. The assumption A3 is that, in the aggregate, the effect of the intentions of each individual rabbit and fox is random.

These three assumptions yield the simple system of ordinary differential equations given by

$$\begin{cases} R' &= aR - bRF, \\ F' &= mRF - nF. \end{cases}$$

The right hand sides are separable, a fact exploited in the analysis of the system, and thus can be written as

$$\begin{cases} R' &= R(a - bF), \\ F' &= F(mR - n). \end{cases}$$

The model can be calibrated in a similar way as Richardson's model in Chapter 1. Given population levels, derivatives are estimated by differences and plotted as R'/R against F. This is followed by a least squares linear fit for for a, b and analogously F'/F against R for m, n.

Remark: There are many possible models. *These equations can be modified to incorporate different types of population growth. For instance, if the prey population undergoes a* logistic *growth process in the absence of the predators the equations are modified as follows:*

$$\begin{cases} R' &= aR(1 - R/K) - bRF, \\ F' &= mRF - nF, \end{cases}$$

where K represents the carrying capacity of the rabbit population. Another possibility is to modify the predator's equation with a carrying capacity as follows. If it takes J prey per unit time to support one predator then we can regard x/J as the carrying capacity of predators. Inserting this for the carrying capacity in the logistic equation changes the predators equation to give the system:

$$\begin{cases} R' &= aR - bRF, \\ F' &= c\left(1 - \frac{F}{(R/J)}\right)F. \end{cases}$$

Finally, the prey population can also be modelled by logistic growth where its carrying capacity is reduced by the number of predators

$$\text{capacity} = K - bF.$$

This gives a coupled logistic model

$$\begin{cases} R' &= aR\left(1 - \frac{R}{K-bF}\right), \\ F' &= c\left(1 - \frac{F}{(R/J)}\right)F. \end{cases}$$

With so many plausible models the important questions are naturally:

- *Which make accurate predictions?*

  and

- *Which are useful for understanding qualitative interactions?*

5.3   Analysis of the Predator-Prey Model

Consider the predator-prey model. Renaming the variables x and y (from R and F, respectively) to make the analysis proceed more familiarly, gives the system

$$\begin{cases} x' &= x(a - by), \\ y' &= y(mx - n), \end{cases}$$

where a, b, m, n are positive parameters. To sketch the phase portrait of this model we first find the nullclines. Note that

$$\begin{array}{llll} x' = 0 & \text{if and only if} & x = 0 & \text{or} \quad y = \frac{a}{b}. \\ y' = 0 & \text{if and only if} & y = 0 & \text{or} \quad x = \frac{n}{m}. \end{array}$$

The nullclines are thus four lines $x = 0, y = 0, x = \frac{n}{m}, y = \frac{a}{b}$ with two equilibria $(0,0)$ and $\left(\frac{n}{m}, \frac{a}{b}\right)$, Figure 5.2.

Trajectory directions are easily found. From $x' = x(a - by)$ it follows that $x(t)$ is increasing (the trajectories move to the right) if and only if $y < a/b$,i.e., below the horizontal line. Trajectories move to the left above the horizontal line. Similarly, $y(t)$ is increasing when x is to the right of the vertical line $x = n/m$

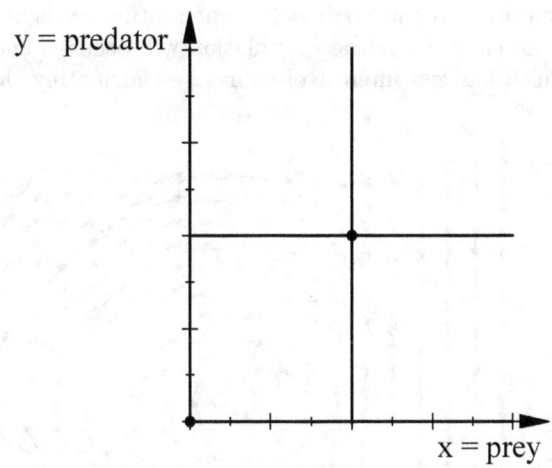

Figure 5.2:   The Lotka-Volterra model has nullclines $x = 0$, $y = 0$, $x = \frac{n}{m}$ & $y = \frac{a}{b}$ and equilibra $(0,0)$ and $\left(\frac{n}{m}, \frac{a}{b}\right)$.

and decreasing to the left of it. This is depicted in the next two figures of the nullclines and trajectory directions.

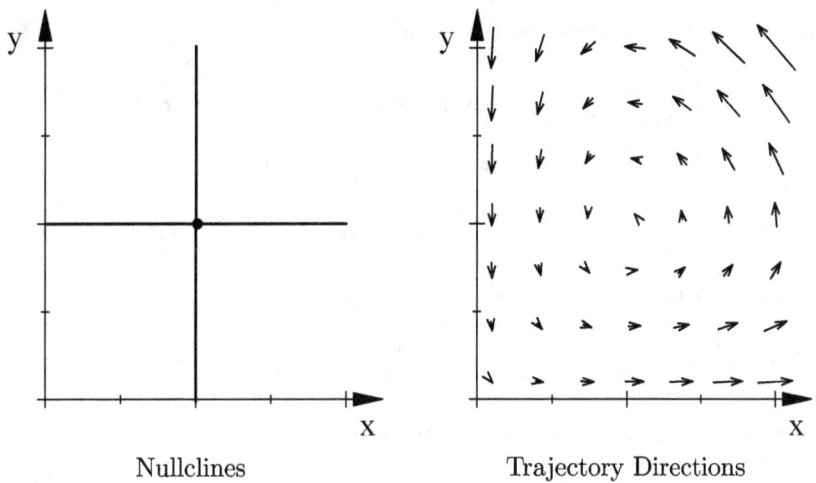

Nullclines   Trajectory Directions

A computer plot of a denser slope field is next in Figure 5.3. As is often the case, a hand drawn figure can communicate a better understanding of the dynamics than a precise, computer generated plot. The intersection of the lines

$$y = \frac{a}{b} \quad \text{and} \quad x = \frac{n}{m}$$

is the interesting critical point $(n/m, a/b)$, representing levels of the two populations at which no change in either population will occur. (The other critical point is $(0, 0)$ which has less interest of course.)   Calculating the linearization

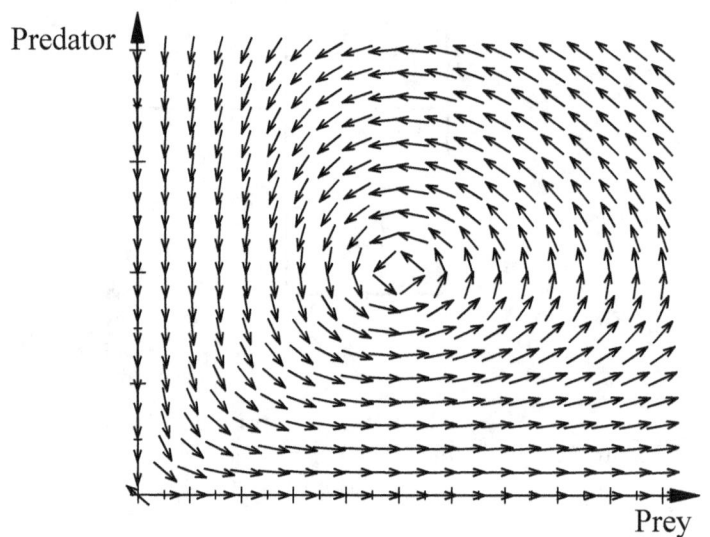

Figure 5.3: Slope Field for the Lotka-Volterra Model

at $(n/m, a/b)$, the linearization matrix is

$$\begin{bmatrix} a - by & -bx \\ my & mx - n \end{bmatrix} \text{ evaluated at } (x, y) = (n/m, a/b)$$

yielding the 2×2 linearization matrix:

$$A = \begin{bmatrix} 0 & -bn/m \\ ma/b & 0 \end{bmatrix}.$$

The characteristic equation of A is found in the usual way:

$$0 = \det(A - \lambda I) = \det \begin{bmatrix} -\lambda & -bn/m \\ ma/b & -\lambda \end{bmatrix} = (-\lambda)^2 - (-bn/m)(ma/b).$$

Thus the characteristic equation is, after simplification,

$$\lambda^2 + an = 0$$

so that

$$\lambda = \pm i\sqrt{an}.$$

The critical point $(n/m, a/b)$ is a center for the linearized problem. Thus, for the *nonlinear* Lotka-Volterra model the critical point may be a center, a stable spiral, or an unstable spiral. A plot of a few trajectories from a numerical routine, below in Figure 5.4, strongly suggests that it is a center for the nonlinear problem. In the next section we will develop new mathematical tools and use

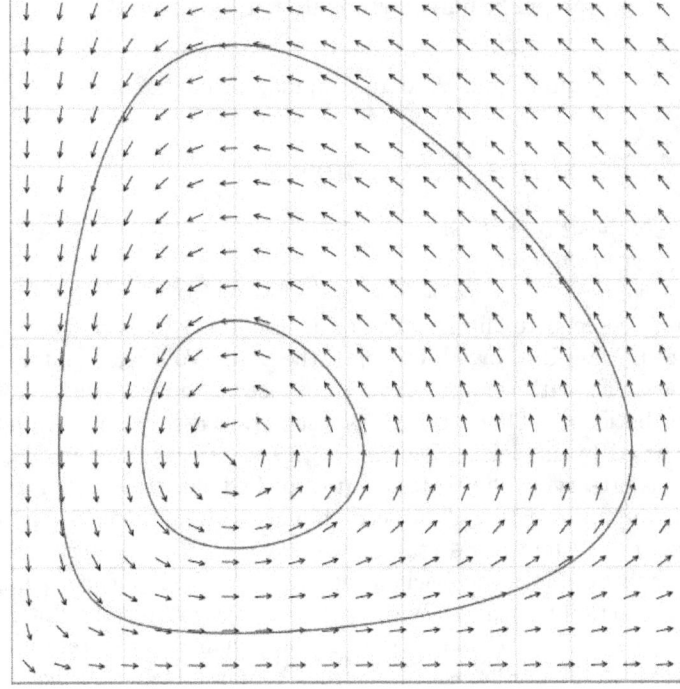

Figure 5.4: Numerical phase plane plot of LV Model

them to show that the critical point is a center for the nonlinear problem.

Remark: Harvesting Modifications. Suppose members of both communities are removed at a rate proportional to their number. The predator-prey model would then be modified to include harvesting terms, such as:

$$\begin{cases} R' &= aR - bRF - cR, \\ F' &= mRF - nF - dF. \end{cases}$$

If we regroup the terms above, we can see the net effect that harvesting has on the predator prey model. The system of differential equations can be rewritten as

$$\begin{cases} R' &= (a-c)R - bRF, \\ F' &= mRF - (n+d)F. \end{cases}$$

It is thus clearly seen that harvesting has the two net effects. It replaces a in the predator-prey model by $a - c$, thus decreasing the growth rate of the prey, and replace n by $n + d$, thus increasing the death rate of the predator. Additionally, harvesting affects the critical point so that, at equilibrium,

$$\left(\frac{n}{m}, \frac{a}{b}\right) \text{ is replaced by } \left(\frac{n}{m-d}, \frac{a-c}{b}\right):$$

equilibrium Prey population *increases*

and

equilibrium Predator population *decreases*.

5.4   Oscillations in models

"Oh, what a difficult task it was.
    To drag out of the marsh
    the hippopotamus" - Korney Ivanovic' Chukovsky

Natural phenomena exhibit many stable oscillatory or periodic patterns yet these cannot be described by, for example, the Lotka-Volterra equations or other dynamical systems with centers. This is because centers are unstable to higher order perturbations . There are two substantive cases where oscillations are stably predicted by models. The first is in conservative systems. Typically, perturbations preserving the "total energy" or first integral frequently preserve oscillatory solution structure. The second, and more mathematically satisfying one, is when the system has a stable limit cycle. This consideration will lead us, in a later section, into the fascinating and beautiful Poincaré-Bendixon theorem. In this section we will study conservative systems and the Lotka-Volterra equations as a specific example thereof. If the dynamically system is linear the most general periodic motion is an ellipse in the phase plane. We have seen that the nonlinear model of predator prey interactions given by the Lotka-Volterra equations:

$$\begin{cases} x' & = & (a-by)x, \\ y' & = & (-n+mx)y, \end{cases} \tag{5.1}$$

where a, b, c, d are positive parameters, has non-elliptical periodic orbits. Let us examine the system (5.1) more closely.

Definition 38 *A set D in the $x-y$ plane is **open** if there is a (small enough) ball about each each point in the set that is entirely in the set:*

$$\text{for all points } (a,b) \ \in \ D \ \text{ there is a radius } r > 0$$
$$\text{such that } \{(x,y) \ : \ (x-a)^2 + (y-b)^2 < r^2\} \subset D.$$

*A set D in the $x-y$ plane is **connected** if any two points inside D can be connected by a continuous curve that lies entirely inside D. A **region** D in the $x-y$ plane is an open connected set. A region D is **simply connected** if it is an open connected set that has no holes[2].*

---

[2] This is not a completely precise definition but it completely expresses the idea. For full precision, having *no holes* means that *any closed curve inside D can be continuously shrunk to a point so as to stay inside D the whole time*. If a region D has a hole, then any curve that circles the hole must cross outside D if it is shrunk to a point.

We can now define a first integral of a nonlinear system. First integrals are mathematical expressions of the idea that the system has some quantity, like an energy, that is exactly conserved.

Definition 39 *Let D be a region in the plane. A real valued function $F(x, y)$ defined on D that is continuous with continuous partial derivatives $\frac{\partial F}{\partial x}, \frac{\partial F}{\partial y}$ is a* **first integral** *of the nonlinear system*

$$\begin{cases} x' & = & P(x, y), \\ y' & = & Q(x, y) \end{cases}$$

on D, if $F(x(t), y(t))$ is constant on any trajectory $(x(t), y(t))$ while that trajectory lies in D.

First integrals are not unique. If $F(x, y)$ is a first integral then so is

$$F(x, y) + c, \; cF(x, y), \; e^{F(x,y)}$$

and so on. If $F(x, y)$ is a first integral then, being constant along a trajectory implies that

$$0 = \frac{d}{dt} F(x(t), y(t)) = F_x(x(t), y(t)) \cdot x'(t) + F_y(x(t), y(t)) \cdot y'(t),$$

or (replacing x' by $P(x, y)$ and y' by $Q(x, y)$)

$$0 = F_x(x, y) \cdot P(x, y) + F_y(x, y) \cdot Q(x, y). \tag{5.2}$$

The equation (5.2) can be used to construct first integrals, but a more common approach that often works when $P(x, y), Q(x, y)$ are separable is to use the equation of trajectories in the phase plane:

$$\frac{dy}{dx} = \frac{Q(x, y)}{P(x, y)}. \tag{5.3}$$

5.4.1 Conservative systems

We start with an example. Consider the linear system

$$\begin{cases} x' & = & -y \\ y' & = & +x. \end{cases}$$

The trajectory equation is a separable differential equation

$$\frac{dy}{dx} = -\frac{x}{y},$$

or equivalently

$$\int -y \, dy = \int x \, dx.$$

Integrating both sides yields

$$-\frac{y^2}{2} = \frac{x^2}{2} + C.$$

Multiplying by 2 and rearranging (with a different constant C)

$$x^2 + y^2 = C \tag{5.4}$$

This equation yields a first integral

$$F(x, y) = x^2 + y^2,$$

which is constant along trajectories. This first integral is defined (and smooth) on the entire plane. Trajectories of $x' = -y, y' = +x$ are level curves of $z = F(x, y)$ $(= x^2 + y^2)$, i.e., circles.

Definition 40 *The system of ordinary differential equations is* **conservative**

$$\begin{cases} x' &= P(x, y) \\ y' &= Q(x, y) \end{cases}$$

if it has a first integral defined on the entire plane. If the system has a first integral in a region that trajectories cannot leave then it is conservative in that region.

It is possible that separating variables gives a first integral that is not globally defined. To be a conservative system all that is required is that there exists one first integral that is defined everywhere (and not that every first integral be defined everywhere). In the last example, the system is conservative because $F(x, y) = x^2 + y^2$ is defined everywhere. However, $F(x, y) = \ln(x^2 + y^2)$ is also a first integral that is undefined at $(0, 0)$ but the system is still conservative.

Example: A system that is not conservative. The last example was conservative while the following, similar example, is not:

$$\begin{cases} x' &= x, \\ y' &= y. \end{cases}$$

The trajectory equation is again a separable differential equation

$$\frac{dy}{dx} = \frac{y}{x}$$

or equivalently

$$\int \frac{dy}{y} = \int \frac{dx}{x}.$$

Integrating gives $\ln y = \ln x + C$. The first integral $F(x, y) = \ln y - \ln x$ is only defined for both x and y positive. It can be rearranged into another first integral that is well defined on a larger set. Taking the exponential of both sides eliminates the logarithms and yields the solutions

$$y = Cx \quad \text{or} \quad \frac{y}{x} = C$$

The first integral $F(x, y) = y/x$ is then not defined for $x = 0$ and thus not defined on all of the plane (although it is constant along trajectories). As a result, this construction of a conservative first integral fails. It can be shown that no first integral of this example can be globally defined.

We can verify that is $F(x, y) = y/x$ indeed constant along trajectories directly (using $x' = x, y' = y$):

$$\frac{d}{dt}\left(\frac{y}{x}\right) = \frac{x\,(y') - (x')\,y}{x^2} = \frac{x\,(y) - (x)\,y}{x^2} = 0.$$

Example: A conservative pendulum like equation. Consider the second order scalar equation, for $g(x)$ a smooth, real valued function,

$$x'' + g(x) = 0.$$

This equation can be written as a first order system of differential equations by defining $y = x'$ giving

$$\begin{cases} x' &= y \\ y' &= -g(x) \end{cases}$$

The trajectory equation is again separable

$$\frac{dy}{dx} = \frac{-g(x)}{y} \quad \text{equivalently} \quad \int y\,dy = \int -g(x)\,dx$$

Now, let G be an antiderivative of g, so that $G'(z) = g(z)$. Integrating gives

$$\frac{y^2}{2} = -G(x) + C \quad \text{or} \quad \frac{y^2}{2} + G(x) = C$$

Thus, we define the first integral

$$F(x, y) = \frac{y^2}{2} + G(x).$$

We calculate that $F(x, y)$ is constant along trajectories:

$$\begin{aligned} \frac{d}{dt}F(x(t), y(t)) &= F_x(x, y) \cdot x' + F_y(x, y) \cdot y' \\ &= G'(x) \cdot x' + y \cdot y' \\ &= g(x) \cdot y + y \cdot (-g(x)) \\ &= 0. \end{aligned}$$

Example: A conservative system with a nonlinear center. Consider

$$\begin{cases} x' &= y^5, \\ y' &= -x^3. \end{cases}$$

The point $(0,0)$ is the only critical point and a globally well defined and smooth first integral is easy to calculate. Indeed, separating variables

$$\frac{dy}{dx} = -\frac{x^3}{y^5} \quad or \quad \int y^5 dy = \int x^3 dx$$

As previously, this implies that

$$F(x,y) = \frac{y^6}{6} + \frac{x^4}{4}$$

is constant along trajectories. Thus

$$F(x,y) = \frac{y^6}{6} + \frac{x^4}{4}$$

(which is defined on the entire plane) is a first integral for this system. Thus the system is conservative. This function $F(x,y)$, plotted below in Figure 5.5, has a local minimum at the critical point $(0,0)$. The level curves of $F(x,y)$ must be

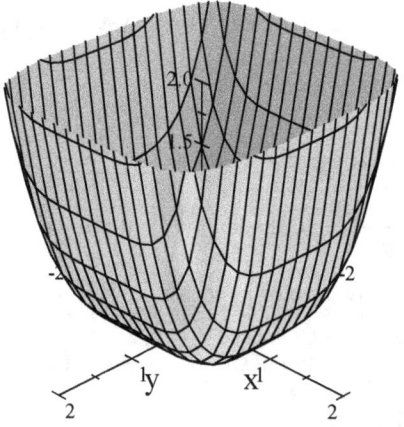

Figure 5.5: The first integral surface $z = \frac{y^6}{6} + \frac{x^4}{4}$

an increasing family of closed curves around $(0,0)$, depicted in Figure 5.6 next.

Since, for a conservative system, trajectories must follow level curves of the first integral, $(0,0)$ must be a center.

5.5   Periodic solutions of the Lotka-Volterra model

Reconsider the Lotka-Volterra system. Equation (5.1) has fixed points $(0,0)$ and $(c/d, a/b)$. $(0,0)$ is a saddle point and $(c/d, a/b)$ is a center for the linearized

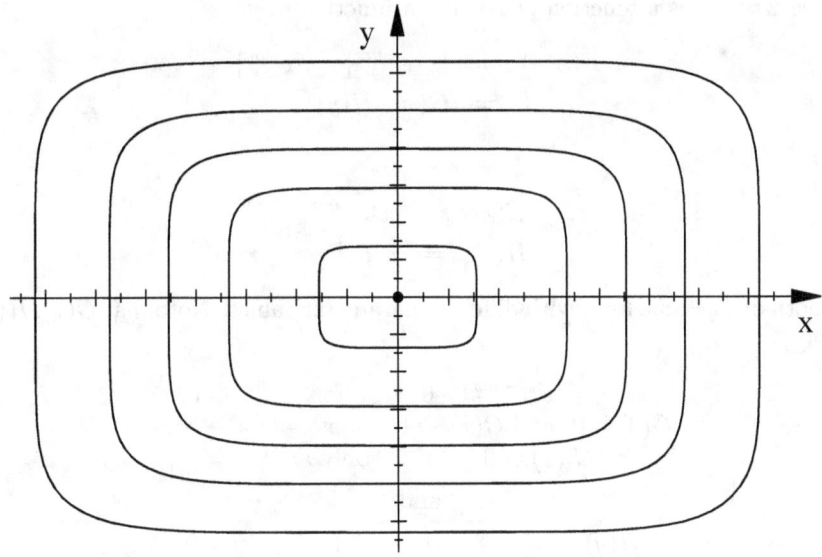

Figure 5.6:
Level curves of $F(x, y)$ depicted:
Trajectories follow levels so $(0, 0)$ is a center

system. To find the first integral of the system (5.1) implies:

$$\frac{dy}{dx} = \frac{(-n + mx)y}{(a - by)x},$$

which is separable. We have

$$\int \frac{a - by}{y} dy = \int \frac{-n + mx}{x} dx.$$

This gives

$$a \ln(y) - by = -n \ln(x) + mx + C.$$

Rearranging

$$\ln (y^a x^n) - (by + mx) = C, \text{ along trajectories.}$$

We remove the singularity of the logarithm by taking $exp(\cdot)$ of both sides. This gives

$$F(x, y) = x^n y^a e^{-mx} e^{-by} \text{ is constant along trajectories.}$$

This first integral $F(x, y)$ is globally defined[3] so the Lotka-Volterra equations is a conservative system. The first integral $F(x, y)$ is also a separable function. It

---

[3]For the parameters a, n integers. $F(x, y)$ is well defined. For noninteger a, n, the exponents in the multiplier $x^n y^a$ cause the usual difficulties for negative population values. (For example if $a = n = 1/2$, $x^n y^a = \sqrt{xy}$.) Since trajectories in the positive quadrant can neither leave it nor enter it from another quadrant, this difficulty is not important. The system is conservative in the positive quadrant and the analysis can proceed.

can be written as a function of x times a function of y:

$$
\begin{aligned}
F(x,y) &= [x^n e^{-mx}] \cdot [y^a e^{-by}] \\
&= G(x) \cdot H(y),
\end{aligned}
$$

where

$$
\begin{aligned}
G(x) &= x^n e^{-mx} \\
H(y) &= y^a e^{-by}
\end{aligned}
$$

are both of the same form with different parameter values. Note that $G(x), H(y)$ satisfy

$$
\begin{aligned}
G(x) = x^n e^{-mx} > 0 \qquad &\text{for} \qquad x > 0, \\
G(0) = 0 \ \text{ and } \ G(x) \to 0 \qquad &\text{as} \qquad x \to \infty, \\
G'(x) = 0 \qquad &\text{only at} \qquad x^* = \tfrac{n}{m},
\end{aligned}
$$

and

$$
\begin{aligned}
H(y) = y^a e^{-by} > 0 \qquad &\text{for} \qquad y > 0, \\
H(0) = 0 \ \text{ and } \ H(y) \to 0 \qquad &\text{as} \qquad y \to \infty, \\
H'(y) = 0 \qquad &\text{only at} \qquad y^* = \tfrac{a}{b}.
\end{aligned}
$$

The functions $G(\cdot)$ and $H(\cdot)$ which are plotted below for typical parameter values.

The maximum of G occurs at n/m and the maximum of H occurs at a/b. Thus, $F(x,y) = G(x) \cdot H(y)$ has a local maximum at the critical point $(n/m, a/b)$. The first integral surface $z = F(x,y) = G(x)H(y)$ is plotted in Figure 5.8 below illustrating this behavior.

Trajectories of (5.1) must stay on level curves of $F(x,y)$. Thus the following holds.

Theorem. *The trajectories of the Lotka-Volterra model must cycle around the critical point. The critical point $(n/m, a/b)$ is a nonlinear center.*

5.6   Harvesting

We have seen the effect of harvesting on equilibrium population levels. We now turn to its effect on time varying populations. Suppose members of both communities are removed at a rate proportional to their number. The predator-prey model is then modified to include harvesting terms:

$$
\left\{
\begin{aligned}
R' &= aR - bRF - cR, \\
F' &= mRF - nF - dF.
\end{aligned}
\right.
$$

If we regroup the terms above, we see the net effect of harvesting. The system of differential equations can be rewritten

$$
\left\{
\begin{aligned}
R' &= (a-c)R - bRF, \\
F' &= mRF - (n+d)F.
\end{aligned}
\right.
$$

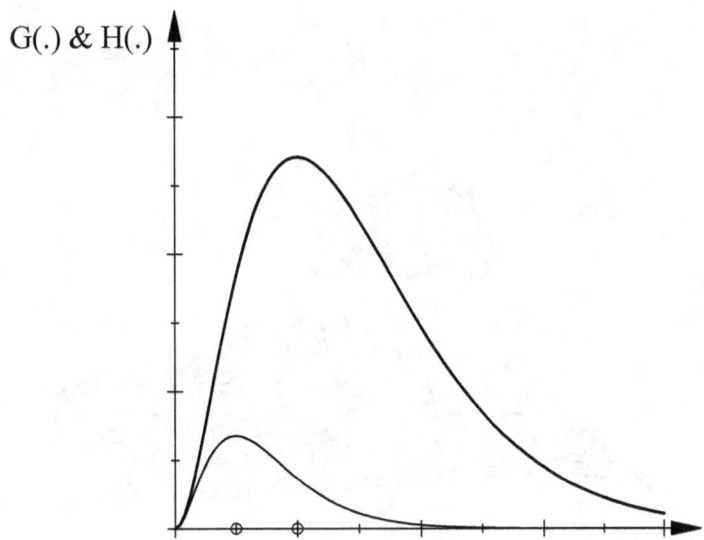

G(.) & H(.)

Typical plots of G($\cdot$) and H($\cdot$)

Figure 5.7:   The first integral of Lotka–Volterra is
$$z = G(x) \cdot H(y)$$

It is thus clear that harvesting has two net effects. It replaces a in the predator-prey model by $a - c$, thus *decreasing the growth rate of the prey*, and replace n by $n + d$, thus *increasing the death rate of the predator*. Additionally, harvesting affects the critical point by

$$\left(\frac{n}{m}, \frac{a}{b} \right) \quad \text{is replaced by} \quad \left(\frac{n}{m - d}, \frac{a - c}{b} \right)$$

The classification of the model's critical point as a center is important for understanding the behavior of the model in a sense we will now make precise. The equilibrium values are the average levels of each population as they oscillate. This is delineated in the following theorem.

Theorem 41 *Consider the Lotka-Volterra system without harvesting ($c = d = 0$)*

$$\begin{aligned} x' &= ax - bxy, \\ y' &= mxy - ny. \end{aligned}$$

Then, the time averages of $x(t)$ and $y(t)$ over one period, T, of $x(t)$ and $y(t)$

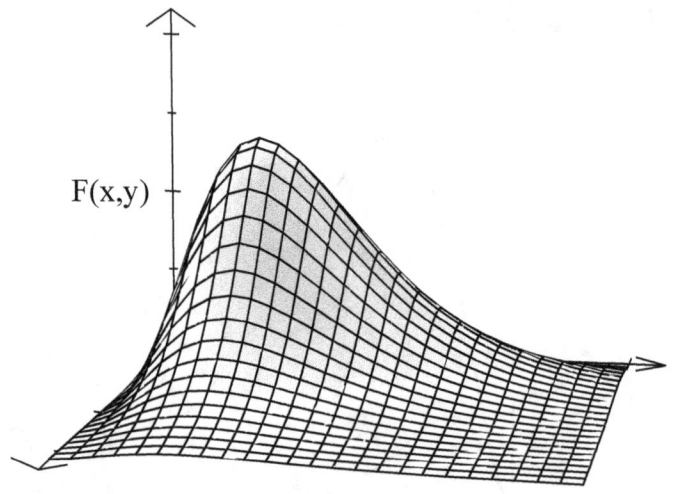

F(x,y)

Figure 5.8:
Lotka-Volterra first integral surface
$$z = F(x, y) = G(x)H(y).$$
It has a local maximum at the equilibrium

are exactly the coordinates of the equilibrium point

$$\frac{1}{T} \int_0^T x(t)\,dt = \frac{c}{d},$$

$$\frac{1}{T} \int_0^T y(t)\,dt = \frac{a}{b}.$$

Proof. The proof exploits, yet again, the fact that the Lotka-Volterra equations are separable. Let the means be denoted

$$\overline{x} = \frac{1}{T} \int_0^T x(t)\,dt,$$

$$\overline{y} = \frac{1}{T} \int_0^T y(t)\,dt.$$

The Lotka-Volterra equations can be rewritten as

$$\frac{x'}{x} = a - by,$$
$$\frac{y'}{y} = mx - n.$$

Since x, y are T−periodic, so are $ln(x), ln(y)$. Further

$$\frac{x'}{x} = \frac{d}{dt}\ln(x)$$

and

$$\frac{y'}{y} = \frac{d}{dt}\ln(y).$$

Thus the equations are

$$\begin{array}{rcl} \frac{d}{dt}\ln(x) & = & a - by, \\ \frac{d}{dt}\ln(y) & = & mx - n. \end{array}$$

We calculate (using T periodicity: $\ln(x(T)) = \ln(x(0))$)

$$0 = \frac{1}{T}\int_0^T \frac{d}{dt}\ln(x)\,dt = \frac{1}{T}\int_0^T a - by\,dt = a - b\overline{y},$$

and

$$0 = \frac{1}{T}\int_0^T \frac{d}{dt}\ln(y)\,dt = \frac{1}{T}\int_0^T mx - n\,dt = m\overline{x} - n.$$

Hence, as claimed,

$$\overline{x} = \frac{n}{m} \quad \text{and} \quad \overline{y} = \frac{a}{b}.$$

■

The critical point is the average values of the species over one period. We have proven that the surprising result observed in the data is determined by the way predators and prey interact. Since, under harvesting

$$\left(\frac{n}{m}, \frac{a}{b}\right) \quad \text{is replaced by} \quad \left(\frac{n}{m-d}, \frac{a-c}{b}\right),$$

we confirm the surprising result, called by Volterra *the law of disturbance of averages*, that:

> *the average predator population is decreased and*
> *the average prey population is increased.*

5.6.1   The remarkable success of the Lotka-Volterra model

For the first mystery of shark populations in the introduction, the Lotka-Volterra model both predicts and explains the observed phenomena: *decrease in harvesting (of prey) increases shark (predator) populations*. If the predator consists of ladybug and the prey is aphids then spraying with a broad spectrum pesticide (increasing harvesting) should have the net effect of increasing the aphid (prey) population while decreasing the ladybug (predator) population. Thus, for the second mystery, *increase in harvesting, increases aphid populations and decreases ladybug populations*. This prediction was borne out when DDT was introduced to kill harmful insects. In particular, one specific example was the case of the cottony cushion insect, which arrived in 1868 from Australia and almost destroyed the U. S. Citrus industry. Its natural predator, the ladybird beetle, was then imported to control it. When DDT was introduced into farming practice, hoping to further reduce the population of cottony cushion insects, the overall effect was to increase the population of the cottony cushion insect and decrease that of the lady bird beetle. The Lotka-Volterra model is thus remarkably successful at explaining mysterious aggregate behavior of complex food chains.

However, the Lotka-Volterra model *does not resolve the third mystery* of a cycle in hare-lynx populations with a *persistent and stable period*. The periods of the predator prey oscillations are not stable but vary with each cycle. Thus, any perturbation of the system would move the oscillation from one cycle to one with a different period.

There is also an inevitable conflict between model comprehensibility and model accuracy in complex applications[4]. As understanding of more details is sought, model complexity inevitably increases. Thus, Lotka-Volterra is a first (successful) step in a hierarchy of models that increase in complexity and in accuracy. We shall develop in later chapters how several of these next steps are developed.

5.7   Models of Interactions of Two Groups

"Interesting phenomena occur when two or more rhythmic patterns are combined, and these phenomena illustrate very aptly the enrichment of information that occurs when one description is combined with another." - Gregory Bateson

"The social nature of hominids engenders cooperation. Groups of hunters working together are able to bring down large dangerous prey animals and are able to defend them from other powerful predators and scavengers. Communication, the transfer of information, is a key capability that strengthens cooperation. It is not a uniquely human trait. For example, elephants communicate and even exhibit cultural memory, as do Japanese macaques. Humans have taken this trait to extraordinary levels. Nevertheless, cooperation is only half of the story. The other half is competition. As cooperation advanced into larger and larger groups, from families to clans and then tribes, competition developed between these groupings. Competition for prey, for territory and for status creates an outlet for that innate ferocity and leads to strife, even armed combat. The history of modern humans is so rife with strife that if it weren't for a very strong desire for cooperatively the record would be one of endless war." - Ron Fox, from Biological Imperatives for Humanity, 2011

Many have been inspired by the Lotka-Volterra model to consider it as a possible model of the interaction of two groups. This has great interest when

---

[4]For example, the Lotka-Volterra model *does not resolve the third mystery* of a persistent and stable 10 year cycle in hare-lynx populations, Gilpin [1973]. Persistent cycles will require further modelling tools.

two language or cultural groups come into contact. It is also a current question in various science fiction movies:

> *Will the zombies or the humans win?*
> *When humans encounter alien civilizations what can then happen?*

To see the connection with Lotka-Volterra models we (briefly) review their derivation. Lotka-Volterra type models are based on two assumptions:

Assumption 1: *If the groups do not contact each other, the group will either grow or decay.* This is expressed as the population undergoing respectively a birth or death process (B or D).

Assumption 2: *The two groups interactions can either help one group grow (C = Cooperation) or suppress its growth (K = Competition).*

For example, the Lotka-Volterra model of Rabbits and Foxes is

$$R' = +aR - bRF,$$
$$\text{and}$$
$$F' = -mF + nRF.$$

This corresponds to:

- $B - D$ (rabbits grow, $+aR$, and foxes die, $-mF$, in the absence of the other), and

- $K - C$ (foxes suppress the growth of the rabbit population, $-bRF$, while rabbits increase the growth of the fox population, $+nRF$).

Within this framework there are at most 16 possible models: one choice of 4 from Table 1

$x \Downarrow \ \searrow \ y \Rightarrow$	Birth	Death
Birth	*B-B*	*B-D*
Death	*D-B*	*D-D*

Table 1

and one choice of 4 from Table 2:

$x \Downarrow \searrow \ y \Rightarrow$	Cooperate	Kompete
Cooperate	*C-C*	*C-K*
Kompete	*K-C*	*K-K*

Table 2

This classification of possible models in terms of cooperation-competition appears in various places. For example, it is (changing C/K to +/-) Odum's +/- scheme for species interactions from his 1953 book. There are only 10 possible models if we eliminate cases where reversing[5] the roles of x and y yield

---

[5] Of course if x = human population and y = zombie population we may wish not to switch the variables.

the same model. Many of these choices have special names; we review a few next. These 16 possible interactions can be expanded to 64 if cooperation and competition inside each group is added.

Example: The choice $B - B$, $K - K$ is **the competing hunters model** given by

$$\begin{cases} x' &= +ax - bxy, \\ y' &= +my - nxy. \end{cases}$$

This model describes two species with a common food source so the predators are in competition with one another. The assumptions are as follows.

Assumption A1. *In the absence of the other, each would undergo a pure birth process.*

Assumption A2. *The competition between the two depresses the rate of growth of each and is proportional to the product of their populations.*

Assumption A3. *There is an infinite supply of prey.*

This model has one positive equilibrium at $\left(\frac{n}{m}, \frac{a}{b}\right)$. The linearization matrix there is

$$\begin{bmatrix} 0 & -\frac{bn}{m} \\ -\frac{am}{b} & 0 \end{bmatrix}$$

with eigenvalues

$$\lambda = \pm\sqrt{an}.$$

Thus $\left(\frac{n}{m}, \frac{a}{b}\right)$ is an unstable saddle point. The phase portrait, easily plotted, shows that one population or the other must vanish as $t \to \infty$. This is known as *the principle of competitive exclusion*.

This competing hunters model has recently been adapted as a possible explanation for the disappearance of Neanderthals from Europe. (There are thousands of explanations given for this. This mathematical cartoon is the 1001st.) The idea was that both populations were competing for the same ecological niche. However, the modern humans population were constantly being augmented by a slow migration out of Africa into Europe. This changes the competing hunters model to

$$\begin{cases} N' &= +aN - bMN, \\ M' &= +mM - nMN + \varepsilon, \end{cases}$$

where $\varepsilon > 0$ represents a small but persistent immigration rate.

Example: The choice D-D, K-K is a model of **combat in guerilla war**. Two armies engage in a guerilla war have numbers denoted $x(t), y(t)$ respectively. They are assumed to satisfy the following.

Assumption A1. *Each side is depleted through accidents, desertions, and so on at a rate proportional to their numbers.*

Assumption A2. *Each side is assumed to suffer losses proportional to the number of contacts between the two.*

These two assumptions lead to the model:

$$x' = -ax - bxy,$$
$$\text{and}$$
$$y' = -my - nxy.$$

The resulting phase plane predicts solutions

$$x(t) \to 0 \text{ and } y(t) \to 0.$$

Example: Social Trends. Social trends can often be modelled by interacting population models where the variable represent those following the trend and those not (yet). For social trends however, the simple interaction model is not adequate because humans do not want to be either isolated or crowded. (If "everyone is doing it" or if no one is doing it then "it" is not desirable, whatever "it" is.) For this reason the response to the number of interactions, $x(t) \times y(t)$, tends to resemble the following, Figure 5.9.

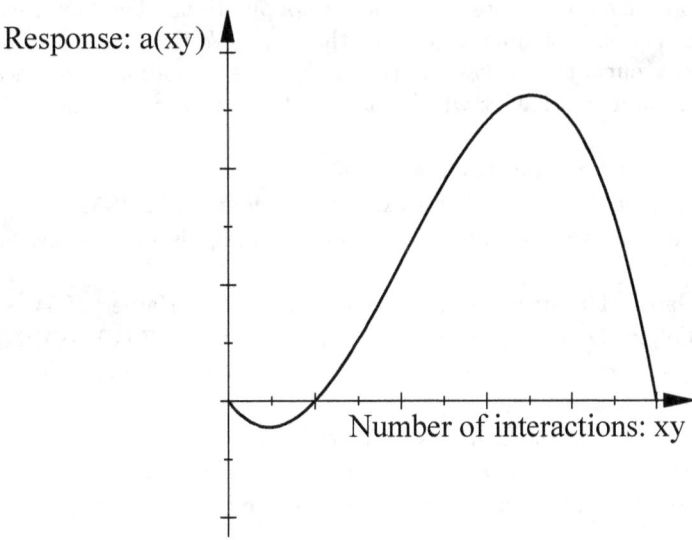

Figure 5.9:   Response to interactions for social trends: Trends disappear when too many or too few follow them.

5.8   A few last thoughts for Chapter 5

"To test a recently developed predator-prey model against reality, I chose the well-known Canadian hare-lynx system. ... The correlation between the model

*and the empirical data gives some idea about the general worth of the model.....
But the regression fit was equally poor. In fact it was worse than poor; it was
impossibly bad." -M.E. Gilpin*

In the modeling process, it is normal to focus on the deficiencies of any
model, preparing its next improvement. This process is possible when model
development is done with assumptions explicitly and precisely made. Testing
models against actual events alerts us to places where our understanding of a
phenomena is incomplete and shows us where greater understanding or better
data is needed. "Model failure" then leads to subsequent success.

Focusing only on model failure also has the danger of completely overlook-
ing the accomplishments of any particular model. We have seen in a previous
chapter that the Malthus model of human population has serious deficiencies.
Nevertheless, it has played a huge role on in the history of human ideas and is the
basis for political, social and economic ideas that still have influence today. The
logistic model provided an improvement in predictability of a population but
has had very limited impact on the history of human ideas. By contrast, the impact of the Lotka–Volterra model has been huge. First, it explained with a very
simple model some mysteries of interacting populations. The biological control
of pests has been a shining success of the Lotka-Volterra model. For human
population groups, the Lotka-Volterra model gave a more accurate description
of complex human behavior which contains both Malthusian competition and
cooperation.

References for Chapter 5

U. D'Ancona, The struggle for existence, Leiden, Brill, 1954.

T.C. Emmell, An introduction to ecology and population biology, New York,
Norton, 1973.

G.F. Gause, The struggle for existence, New York, Hafner, 1934.

M.E. Gilpin, Do hares eat lynx?, American Nat. v.107 (1973) 727-730.

F. Hoppensteadt, Getting started in mathematical biology, Notices of the
AMS, Sept. (1995)969-975.

F.W. Lancaster, Mathematics in Warfare, p. 2138-2157 in: The World of
Mathematics, J.R. Newman editor. Simon & Schuster, 1956.

A.J. Lotka, A new conception of the universe, Harpers magazine, March
1920, pages 477 - 487.

A.J. Lotka, Elements of Physical Biology, Williams and Wilkins, Baltimore,
1925.

D.A. MacLulich, Fluctuations in the numbers of the varying hare (Lepus
americanus). University of Toronto Studies Biological Series 43. University of
Toronto Press, Toronto, 1937.

I. Mahon, Easter Island: The economics of population dynamics and sustainable development in the Pacific context, 113-119 in: Easter Island in Pacific
context: South Seas Symposium (C.M. Stevenson et al., editors), Easter Island
Foundation, Los Osos, 1998.

J.D. Murray, Mathematical Biology, Springer, Berlin,1993.

E.P. Odum, Fundamentals of Ecology, W.B. Saunders, London, 1953.

R. Pearl, Some biological considerations about war, American Journal of Sociology, Volume 46 (1941), pages 487 - 503.

L.B. Slobodkin, Growth and regulation in animal populations, New York, Holt, Rheinehart and Winston, 1961.

L.B. Slobodkin, Comments from a biologist to a mathematician, pages 318-329 in: Ecosystem analysis and prediction, (S. A. Levine, editor), SIAM, Philadelphia, 1975.

V. Volterra, Lecons sur la Theorie Mathematique de la Lutte pour la Vie, Gauthier-Villars, Paris, 1931.

5.9 Exercises for Chapter 5

1. Consider the predator-prey model:

$$\begin{cases} x' &=& x(5 - x - y) \\ y' &=& y(x - 2). \end{cases}$$

(a) What assumptions lead to this system?

(b) Show that the critical points are $(0,0)$, $(5,0)$ and $(2,3)$. Investigate, either numerically or analytically, the stability of the critical point $(2,3)$.

(c) Sketch the phase portrait of this system and compare it to the phase portrait arising from the Lotka- Volterra equations. What do you conclude?

2. Consider the ecological model:

$$\begin{cases} x' &=& x(x - y - 2) \\ y' &=& y(x + y - 4). \end{cases}$$

What assumptions lead to this model? Sketch its phase portrait, either analytically or computationally. Verify that there are two regions in the first quadrant which satisfy the following. If $(x(0), y(0))$ is in Region 1 then $x(t)$ approaches zero and $y(t)$ approaches zero as $t \to \infty$. If $(x(0), y(0))$ is in Region 2 then $x(t)$ approaches zero and $y(t)$ approaches infinity as $t \to \infty$. Find these two regions. Interpret your results from part (c).

3. Consider the ecological model:

$$\begin{cases} x' &=& x - xy + \epsilon x(1 - x) \\ y' &=& -y + xy. \end{cases}$$

Here ϵ is a small parameter; if $\epsilon = 0$ this model reduces to the Lotka-Volterra system. Sketch its phase portrait, numerically or analytically, for different small values of ϵ and observe the behavior of periodic trajectories in this system. Describe your observations. Can you draw conclusions about the stability of periodic orbits?

4. Suppose the assumptions behind the Lotka-Volterra predator-prey model are modified by assuming the prey undergoes logistic growth in the absence of the predator. Develop the resultant system of ordinary differential equations. Pick some parameter values and sketch the resulting phase portrait. Compare it to the one arising from the Lotka-Volterra system. Do the populations cycle?

5. Repeat the last problem where you, instead, modify the assumptions underlying the predator. Suppose instead that the predator undergoes a logistic growth process with carrying capacity proportional to the number of prey.

6. Write down a dynamic model of the following. Two populations interact in a cooperative manner. Assume:

   A1. *Each would undergo a pure death process in the absence of the other.*

   A2 *Interactions between the two increases the growth rate of each in proportion to the number of interactions.*

   A3. *Crowding of each with other members of the same group decreases the growth rate in proportion to the number of interactions.*

7. Two populations $x(t), y(t)$ interact as described below in words. Find the associated model.

$$\text{Rate of increase of } x = \text{logistic birth process } +$$
$$\text{pure death process } + \text{ crowding reduces growth rate}$$
$$\text{Rate of increase of } y = \text{pure death process } +$$
$$y \text{ eating } x \text{ enhances reproduction of } y + \text{ crowding reduces growth rate.}$$

8. Consider the Lotka-Volterra model. Add to the model the assumption that the prey is harvested at a rate proportion to its population.

   (a) Write down the modified model.

   (b) Take all rate constants to be 1 except the harvesting rate which is taken to be 1/2. Sketch the phase portrait. Now take harvesting rate constant > 1 and repeat. Describe the difference in solution behavior.

9. The following system is a model of Parasite-Host interactions.

   (a) Write down the assumptions on which the model is based.

   (b) Take $c = 0$ and all other parameters to be 1. Develop the phase portrait of the model and describe the model's predictions.

$$H' = aH - bHP - cP^2,$$
$$P' = -mP + nHP.$$

10. Consider the system below for p an odd integer larger then 1^6. Show that

$$F(x,y) = x^2 + y^2 - \frac{2}{p+1}x^{p+1}$$

is a first integral. Using it show that $(0,0)$ is a center for both the linear and nonlinear system:

$$x' = -y, \quad y' = x - x^p.$$

11. Consider the equation
$$z'' + (z')^3 + z = 0.$$

(a) Write it as a first order system by the usual means of $x = z, y = z'$.

(b) $(0,0)$ is an equilibrium for the system. Analyze its stability.

---

^{6}To the experienced this means take $p = 3$ to get started. Work the problem out then go through the solution replacing "3" by "p" and see if the steps are still OK.

Chapter 6

Modeling Epidemics

6.1 Introduction

"Epidemics follow patterns because diseases follow patterns. Viruses spread; they reproduce; they die." - Jill Lepore.

There are two broad types of infectious diseases:

1. **Micro-parasitic diseases**. These are caused by bacteria and viruses. Micro parasitic diseases reproduce inside each host and are transmitted directly from host to host. Examples of this type include chicken pox, flu, smallpox and many others.

2. **Macro-parasitic diseases**. These are due to worms and protozoans, have typically a more complicated life cycle and often involve transmission by a carrier. Examples of this type include schistosomiasis (a worm) and malaria (a protozoan).

This chapter studies a successful approach for modeling the spread of micro parasitic diseases. One of the first attempts, which proved unsuccessful, was to model the disease as two populations: parasite $P(t)$ and host $H(t)$. This leads to a predator prey system, such as:

$$\begin{cases} H' &= aH - bHP, \\[2mm] P' &= -mP + nPH. \end{cases} \tag{6.1}$$

Unfortunately, the predator-prey model does not capture the essential feature of epidemics: that they are spread through contact between infected hosts and uninfected hosts. It is not surprising therefore that (6.1) does not agree with observed data from epidemics. We shall develop herein the successful *SIR* model of epidemics due to Kermack and McKendrick in 1927. This model involves tracking the evolution of the following populations:

$S(t) :=$ the number susceptible to infection,

$I(t) :=$ the number of infecteds,
$L(t) :=$ the number of latently infected, and
$R(t) :=$ the number of removed from the epidemic.
Before developing the SIR model let us first review some examples of epidemics in history.

6.1.1   Epidemics in History

"In spite of the advances of medicine, deathly epidemics are more menacing than ever before." - *Christian de Duve*

Epidemics have played a determining role in human history:

- In the middle of the bronze age in Ancient Egypt plagues of smallpox, tuberculosis, leprosy and dyptheria were well known.

- 430 B.C. In 430 B. C. a devastating plague swept through Athens, killing between thirty and sixty percent of the people in the city. This plague had the effect of essentially deciding the Peloponnesian wars in Sparta's favor and ending the Periclean golden era.

- Later, plague, measles and smallpox depopulated the Roman empire contributing to its collapse.

- 1400. In the fifteenth century, bubonic plague, known then as *the Black Death*, killed 25 million out of the 100 million people in Europe. Many of these died within the first 5 years of its introduction.

- 1520. In 1520 the Aztecs, due to a plague of small pox, saw half of their 3.5 million people die. During this century, small pox killed 13 million in Mesoamerica.

- 1660. Bubonic plague was extensive in England.

- 1760. The first mathematical model of a smallpox epidemic was developed by the great mathematician Daniel Bernoulli (1700-1782).

- 1875. In 1875 the king of Fiji and his son visited Australia, bringing back measles. Over 40,000 of 150,000 Fijians died.

- 1905. Plague hits Bombay (now Mumbai), India. Extensive records were kept and are available to test models.

- 1918-1921. From 1918 to 1921, 25 million people in the Soviet Union contracted typhus and one in ten died.

- 1919. In 1919, 60-100 million people died in a worldwide influenza outbreak. In 1957 and 1963 influenza pandemics killed two and one million people respectively.

- 1927. Kermack and McKendrick gave the first successful epidemic model. Their model's predictions gave a nearly exact fit for the Bombay data.

- 1978 - present. The Center for Disease Control (CDC) estimates that the U.S. AIDS/HIV epidemic began around 1978. In 1983 the virus was first identified. By the year 2000, 774,467 cases of AIDS were reported in the U.S.. The UNAIDS program estimated that by December 2010 the AIDS epidemic had reached 34 million people, including 3.4 million children.

- 2010. Concerns include Bird flu, Zika, Ebola and SARS. New viruses are exposed by melting arctic ice. Labs are also actively trying to construct yet more toxic viruses.

- Epidemic models are also being used to understand other things that spread through contact like wars, obesity, pop star fandom (such as "Bieber fever"), drug addiction and domestic violence.

- 2020. The Covid pandemic has caused the greatest social and economic disruption worldwide since the great depression. From 2019-August, 2020 it caused 25 million deaths making it the 5th deadliest pandemic in human history.

6.2 The SIR model

Infectious diseases are quite varied and spread through many different mechanisms. Thus, no one mathematical model can capture the central features of all epidemics. The SIR model studies those infectious diseases that progress through the following stages.

1. *Initial infection .*

2. *Latent period: No symptoms have appeared and communication of the disease is not possible.*

3. *Infectious period.*

4. *Death or recovery.*

5. *If recovery than either:*

    (a) *permanent immunity,*

    (b) *temporary immunity, or*

    (c) *disease carrier.*

The basic issue in modeling epidemics, first set forth in 1927 by W.O. Kermack and A.G. McKendrick, is to estimate the number of people who will be affected by the epidemic. The population is divided into the four groups.

Definition 42 *The susceptibles,*

$S(t) :=$ *the number of people in the population who are currently uninfected and are susceptible to infection.*

The latently infected,

$L(t) :=$ *the number of people in the population who are currently infected but not infectious.*

The infected,

$I(t) :=$ *the number of people in the population who can currently transmit the infection.*

The removed,

$R(t) :=$ *the number of people in the population who have had the disease and cannot infect others, either through death, isolation or permanent immunity.*

The first simple model we study is based upon the following assumptions.

Assumption A1. *The epidemic spreads at a much faster rate than the population increases. In other words, for the model the total population can be assumed fixed: There is a constant population size N with*

$$S(t) + L(t) + I(t) + R(t) = N.$$

Assumption A2. *The rate of spread of the epidemic is proportional to the rate of contact between susceptibles and infecteds. In other words, $S(t)$ satisfies the following equation.*

$$S' = -bIS, \quad \text{for all } t > 0.$$

*Here $b > 0$ is a constant representing the **infection rate**.*

A closure assumption is now needed regarding another variable, either L or R. For simplicity we shall assume that this epidemic has no latency period, such as occurs with typhus fever, spread by the bite of a hair louse.

Assumption A3. *The particular epidemic studied has no latency period:*

$$L(t) = 0.$$

Assumption A4. *Individuals are removed from the infected at a rate proportional to the number of infected. In other words $R(t)$ satisfies the following equation.*

$$R' = rI, \quad \text{for all } t > 0,$$

*where r is a nonnegative constant representing the **removal rate**.*

Assumptions 1, 2, 3 and 4 lead to the famous SIR model, given by the following system of three equations:

$$\begin{cases} S' &= -bIS \\ R' &= rI \\ S(t) + I(t) + R(t) &= N. \end{cases} \tag{6.2}$$

Here $t > 0$, $S(0)$ and $I(0)$ are known, and N is the fixed population size. When unknown, the initial conditions is commonly taken to be

$$I(0) = 1 \quad \text{and} \quad S(0) = N - 1.$$

6.3 Analysis of the SIR model: No Removals

First we consider the special case of the SI model which assumes no removals, $R(t) = 0$. In this case the equations (6.2) can be solved explicitly as follows. As $R(t) = 0$ for all $t > 0$ we have,

$$S(t) + I(t) = N,$$

a fixed constant. We can solve for $S(t)$ in terms of $I(t)$ and then the SIR model with no removal reduces to a single differential equation

$$I' = bI(N - I).$$

This is a form of the logistic equation with carrying capacity $K = N$ as it can be rewritten as

$$I' = bN\left(1 - \frac{I}{N}\right)I.$$

The logistic equation can be solved exactly by separating variables. With initial condition $I(0) = 1$ its solution is given by

$$I(t) = \frac{N}{1 + (N-1)e^{-bNt}}.$$

Using $S(t) = N - I(t)$ we also have

$$S(t) = \frac{N}{1 + \frac{e^{bNt}}{N-1}}$$

The graphs of $I(t)$ and $S(t)$ are the classic logistic curves in the next two figures.

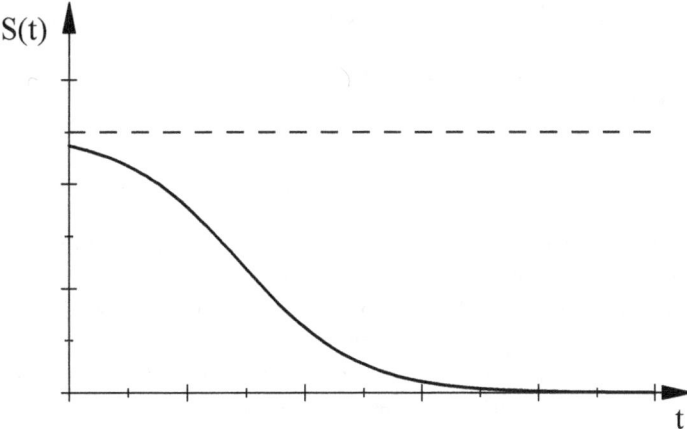

Figure: $S(t)$, **No Removals**

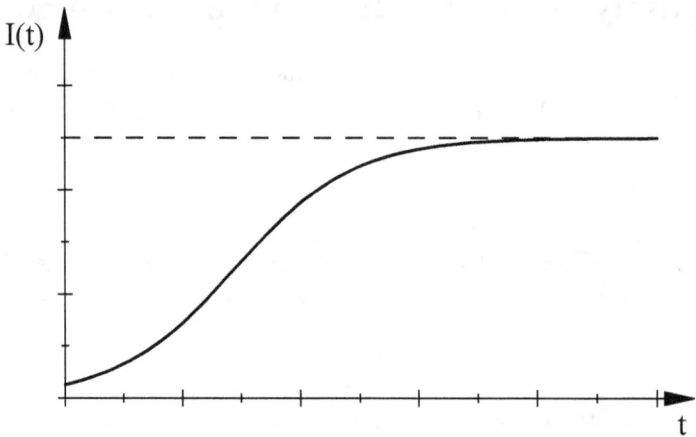

Figure: $I(t)$, **No Removals**

The Epidemic Curve. The basic data collected in an epidemic is the number of new cases each day or week. If T is the time interval for which data is collected, the data represents:

$$\frac{I(t+T) - I(t)}{T},$$

which is approximately $I'(t)$. $I'(t)$ is given by

$$I'(t) = \frac{N^2(N-1)be^{-bNt}}{(1 + (N-1)e^{-bNt})^2}.$$

The graph of the function $I'(t)$ is called **the epidemic curve**, plotted in Figure 6.1 next. Since we have a closed form formula for $I(t)$ it is easy to calculate that

$$t_{\max} = \frac{\ln(N-1)}{bN}, \ I(t_{\max}) = \frac{N}{2} \text{ and } I'(t_{\max}) = \frac{N^2b}{4}.$$

Elementary analysis of the curve shows that

$$I_{\max} = \frac{N}{2} \text{ at } t_{\max} = \frac{\ln(N-1)}{bN}$$

$$and$$

$$I'(t_{\max}) = \frac{N^2b}{4}.$$

The analysis shows that the larger b is the faster the epidemic reaches its peak (t_{max} is smaller) and the smaller b is the slower the epidemic progresses. This parameter b is a measure of the transmissibility of the disease which is affected by density or crowdedness of the population.

If there are no removals, then the SI model predicts that ultimately everyone must eventually catch the disease. This feature does not describe many epidemics. Thus, we must now consider the more complex SIR, model, allowing removals.

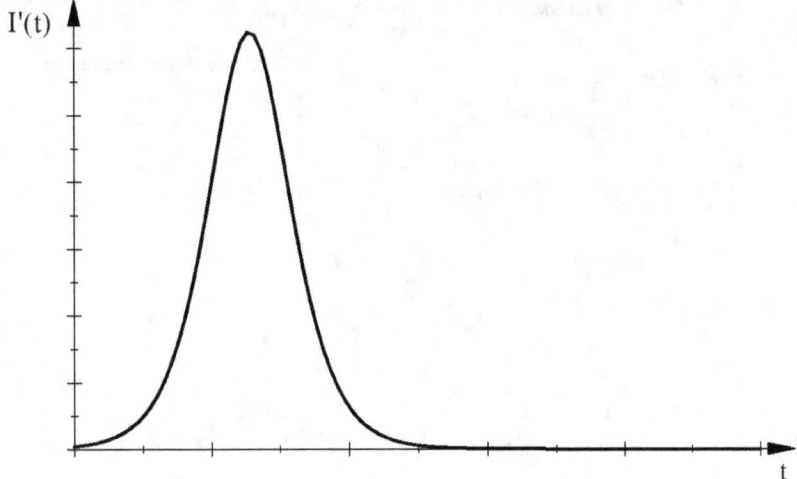

$$\text{The epidemic curve: } I'(t) \text{ vs } t$$

Figure 6.1: $I_{\max} = \frac{N}{2} \& I'_{\max} = \frac{N^2 b}{4}$ at $t_{\max} = \frac{\ln(N-1)}{bN}$

$$\text{where } I(t) = \# \text{ infected,}$$

6.4   The Full SIR Model

To get an idea how fast an epidemic can develop, a few plots of the 2017 flu season (for which the forecasters did not guess the active strains correctly) from the CDC web site[1], https://www.cdc.gov/flu/weekly/usmap.htm, are given next.

Obviously, this was a case when an epidemic occurred. We now analyze the question

When does an epidemic occur?

or equivalently

What are the conditions necessary to prevent an epidemic?

Consider the case when the removal rate r is positive. The *relative removal rate* is the ratio

$$\text{relative removal rate: } p = \frac{r}{b}.$$

---

[1] At https://www.cdc.gov/media/subtopic/questions.htm the CDC states "CDC materials available on the web site are in the public domain (free of copyright restrictions) unless otherwise noted." In addition to appreciation for the central mission of the CDC, I thank the CDC for making graphics such as these available for general use.

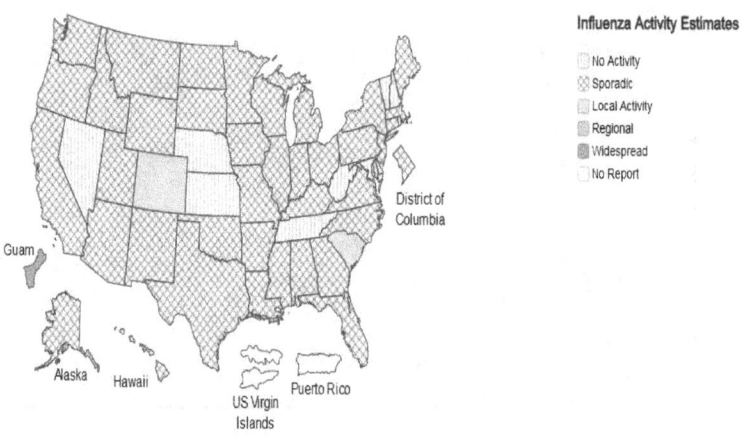

Figure 6.2: Week 40 Flu Map in 2017

We must examine the following coupled system for S, I, and R. Given $t > 0, b > 0$ and $r > 0$ and given known values $S(0), I(0),$ and $R(0)$ we have

$$\begin{cases} S' &= -bSI \\ I' &= bSI - rI \\ R' &= rI \\ S + I + R &= N \end{cases} \tag{6.3}$$

We can eliminate one variable, say R, and consider the 2×2 subsystem by phase plane methods. As $R = N - S - I$,

$$R' = -S' - I'.$$

The third equation of the system (6.3) now gives

$$S' + I' = -rI.$$

Adding the first and second equations of the system (6.3) gives:

$$-rI =: S' + I' = -rI - bSI + bSI = rI.$$

The third and fourth equations of the system (6.3) are eliminated and one can sketch the phase plane of the (S, I) subsystem:

$$\begin{cases} S' &= -bSI \\ I' &= bI(S - r/b). \end{cases}$$

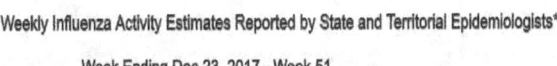

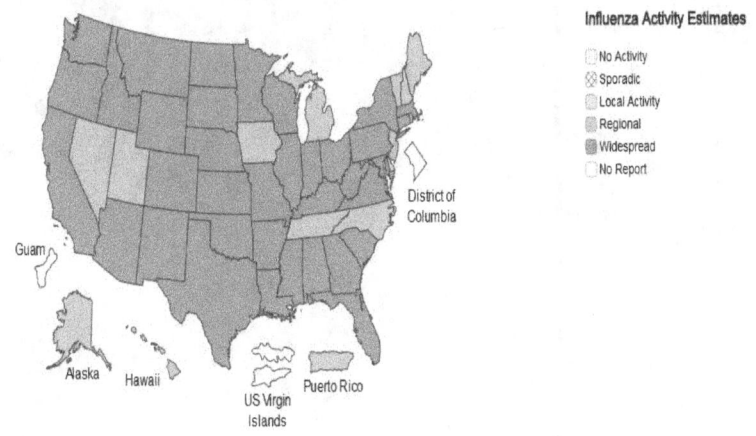

Figure 6.3:
Week 51 Flu Map in 2017:
In only 11 weeks
flu spread across the US.

This yields the SIR model's phase portrait in $S-I$ variables in Figure 6.4 below.

From the above figure, the number of infected $I(t)$ must be strictly decreasing unless $S(0)$ is bigger than $r/b = p$. This is an example of a *threshold phenomenon: there can be no epidemic unless*

$$S(0) > \frac{r}{b}.$$

As a key conclusion of the phase plane analysis, epidemics can be prevented only by one of three measures:

1. *decreasing the susceptibles population, for example, by vaccination,*

2. *increasing the removal rate, for example, by quarantines, or*

3. *decreasing the infection rate, by, for example, better hygiene or less crowdedness.*

Theorem 43 (Threshold Theorem for Epidemics.) *Consider the SIR model (6.3). If*

$$S(0) < \frac{r}{b}$$

then $I(t) \to 0$ monotonically as $t \to \infty$.

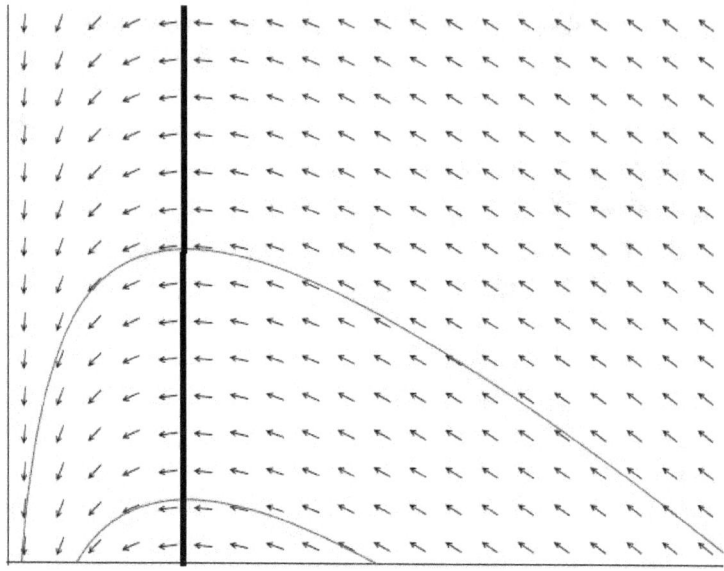

Figure 6.4:
SIR model trajectories
Vertical axis is # infected I
Horizontal axis is # succeptible S

If

$$S(0) > \frac{r}{b}$$

then $I(t)$ first increases as t increases and then decreases monotonically to 0. The $\lim_{t \to \infty} S(t)$ is the unique positive root of the following equation

$$F(x) := S(0)e^{-b/r(N-x)} - x = 0.$$

Proof. That $I(t)$ approaches zero monotonically as t goes to infinity when $S(0)$ is less than r/b is clear from the previous figure. When $S(0)$ is bigger than r/b, that $I(t)$ increases at first is equally clear. $I(t)$ will then decrease monotonically to 0 if there is a T such that $S(T) = r/b$. The trajectory equation in the $S - I$ phase plane is:

$$\frac{dI}{dS} = \frac{I(bS - r)}{-bSI} = -1 + \frac{p}{S}.$$

By separating variables we find:

$$I = -S + p \ln S + C,$$

or

$$I(t) - I(0) = S(t) + p \ln \left(\frac{S(t)}{S(0)} \right), \qquad (6.4)$$

for all times $t > 0$.

As $S(t)$ decreases monotonically and is always non-negative, the limit of $S(t)$ as $t \to \infty$ exists. By equation (6.4) the limit of $I(t)$ as $t \to \infty$ must also exist. Calling these two limits

$$S^* \quad : \quad = \lim_{t \to \infty} S(t)$$

and

$$I^* \quad = \quad \lim_{t \to \infty} I(t).$$

respectively, equation (6.4) implies they satisfy

$$I^* = I(0) + S^* + p \ln \left(\frac{S^*}{S(0)} \right).$$

Further, this implies that (S^*, I^*) must be an equilibrium. Since the only equilibria are on the S axis, I^* must satisfy $I^* = 0$. As $I^* = 0$, S^* is a root of

$$N - S + p \ln \left(\frac{S}{S(0)} \right) = 0,$$

or, equivalently, $F(x) = 0$, where (by rearrangement)

$$F(x) := S(0)e^{-b/r(N-x)} - x.$$

Since $I(t)$ approaches zero as t approaches infinity, $(S(t), I(t))$ must cross the line $S = r/b$, after which $I(t)$ decreases monotonically. Consider now the equation for S^*:

$$F(x) := S(0)e^{-b/r(N-x)} - x = 0. \tag{6.5}$$

This equation must have exactly one positive root by the intermediate value theorem, since $F(0)$ is positive and

$$S(N) = S(0) - N < 0,$$

is negative. This completes the proof. ∎

6.4.1   Estimating R_∞, S_∞

For severe diseases R' is the most reliable data as it represents the death rate. For example, in 1660 Bubonic plague struck the village of Eyam in England. There is no data for the number infected. On the other hand, over the next months the following deaths were carefully recorded.

Month:	*June*	*July*	*Aug.*	*Sept.*	*Oct.*
#deaths	19	56	77	24	14

Bubonic plague deaths in Eyam England

These numbers are more certain than even the total village population. Plotting them in Figure 6.5 yields a definite pattern:

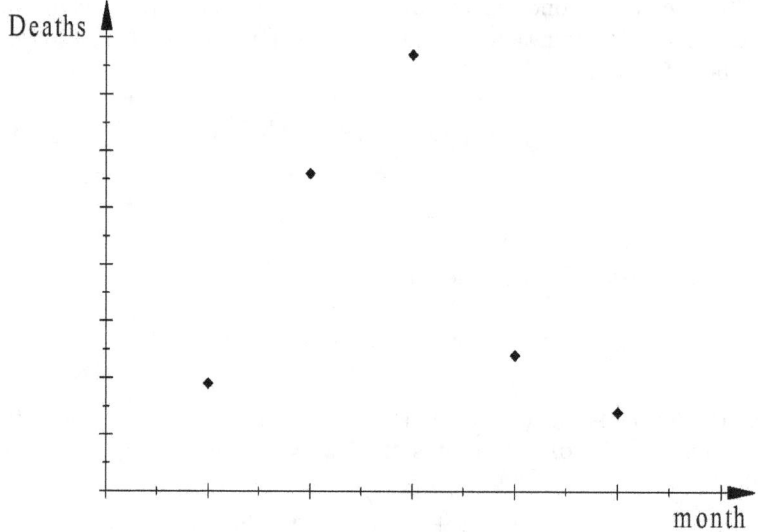

Figure 6.5: Deaths/month during Eyam Plague

Analysis of $R'(t)$ explains this type of pattern. since

$$S' = -bSI$$
$$R' = +rI \qquad \Rightarrow \qquad \frac{dS}{dR} = -\frac{1}{p}S.$$

We can thus analyze the closed equation

$$\frac{dS}{dR} = -\frac{1}{p}S.$$

Separating variables gives an expression for $S(t)$ in terms of $R(t)$

$$S(t) = S(0)e^{-(1/p)R}.$$

Insertion into $\frac{dR}{dt} = r(N - S - R)$ gives a single ordinary differential equation for $R(t)$:

$$\frac{dR}{dt} = r(N - S(0)e^{-(1/p)R} - R). \qquad (6.6)$$

This is not solvable in closed form. Thus, we shall perform a process in applied math sometimes called "negotiating with the problem."

The idea is to replace a problem we cannot solve in closed form with a nearby, close approximation, that can be solved. Normally this is done by linearization (since linear equations can be solved easily). Here a quadratic

Taylor approximation may be used (since quadratic equations can also be solved exactly) as follows. Truncate the Taylor series for e^x

$$e^x = 1 + x + \frac{1}{2}x^2 + \mathcal{O}(x^3)$$

to $1 + x + (1/2)x^2$ so that $e^{-(1/p)R}$ becomes, approximately,

$$e^{-(1/p)R} \simeq 1 - \frac{R}{p} + \frac{1}{2}\left(\frac{R}{p}\right)^2.$$

We then have the approximate model

$$R' = r\left\{N - R - S(0)\left[1 - \frac{R}{p} + \frac{1}{2}\left(\frac{R}{p}\right)^2\right]\right\}.$$

Now let $t \to \infty$. When the epidemic ends $R' = 0$. Letting $t \to \infty$ gives the following equation for $R_\infty = \lim_{t \to \infty} R(t)$:

$$0 = r\left[S(0) - R_\infty - S(0) + S(0)\frac{R_\infty}{p} - \frac{1}{2}S(0)\left(\frac{R_\infty}{p}\right)^2\right].$$

This is a quadratic equation for R_∞ (explaining why the quadratic approximation was used). Solving for R_∞ yields, approximately,

$$R_\infty = 2p\left(1 - \frac{p}{S(0)}\right). \qquad (6.7)$$

Let $S(0) = p + v$ so that

$v = $ *number of susceptibles initially above the threshold p.*

The value of R_∞, S_∞ then become

$$R_\infty \approx \frac{2pv}{p+v},$$

$$S_\infty \approx p + v - R^* - \frac{2pv}{p+v} \approx p - v.$$

Thus the total epidemic is about $2v$ cases, the susceptibles, initially $p + v$ in number, are reduced to about $p - v$ at the end of the epidemic. This key observation was already made by Kermack and McKendrick in 1927! Figure 6.6 below gives the general SIR prediction of $R'(t)$.

Kermack and McKendrick compared the model solution for $R'(t)$ against deaths per week in the Bombay plague epidemic in 1927. Their comparative plot of the model's predictions against the actual data is justifiably the most famous figure in epidemiology. We have replotted below in Figure 6.7 the death/week data from this period and the SIR model's predictions over a 52 week period. The SIR model was very accurate..

R'(t) = deaths/week

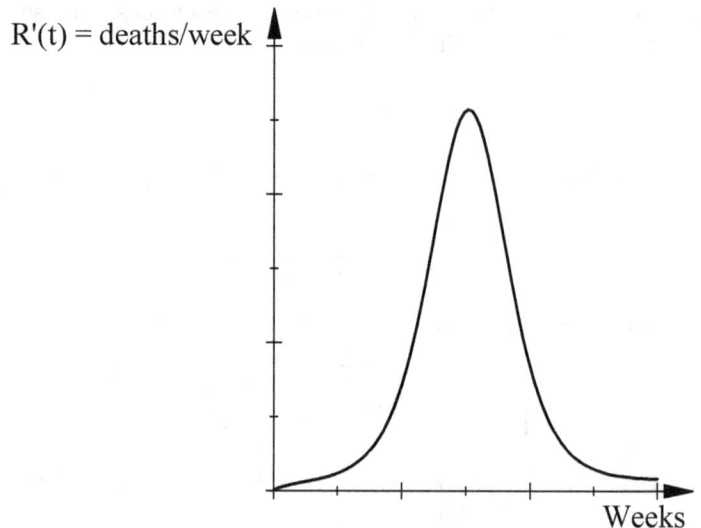

Weeks

$$\text{Figure 6.6:} \quad \begin{array}{l} \# \text{ Deaths per week: } R'(t) \text{ where} \\ R' = r(N - S_0 e^{-(1/p)R} - R) \end{array}$$

6.5   Temporary Immunity, Latency and Asymptomatic Carriers

Temporary Immunity and the SIRS Model. Suppose immunity from the disease is only temporary so that members of the removed group return to become susceptibles at a rate proportional to the number of removed present. This leads to the SIRS model, given by

$$S + I + R = N$$

and:

$$\begin{cases} S' & = & -bSI + gR, \\[2mm] I' & = & bSI - rI, \\[2mm] R' & = & rI - gR. \end{cases} \tag{6.8}$$

The SIRS model allows for the possibility that either the disease is eradicated or the disease will be permanently established in the population.

    Latency and the SEIR Model. The SEIR model includes a latency period (whose population is denoted by $E(t)$). It has been used to model the 1976 and 1995 Ebola outbreaks. The latency group $E(t)$, infected but not infectious, also includes those infecteds who cannot spread the epidemic for any

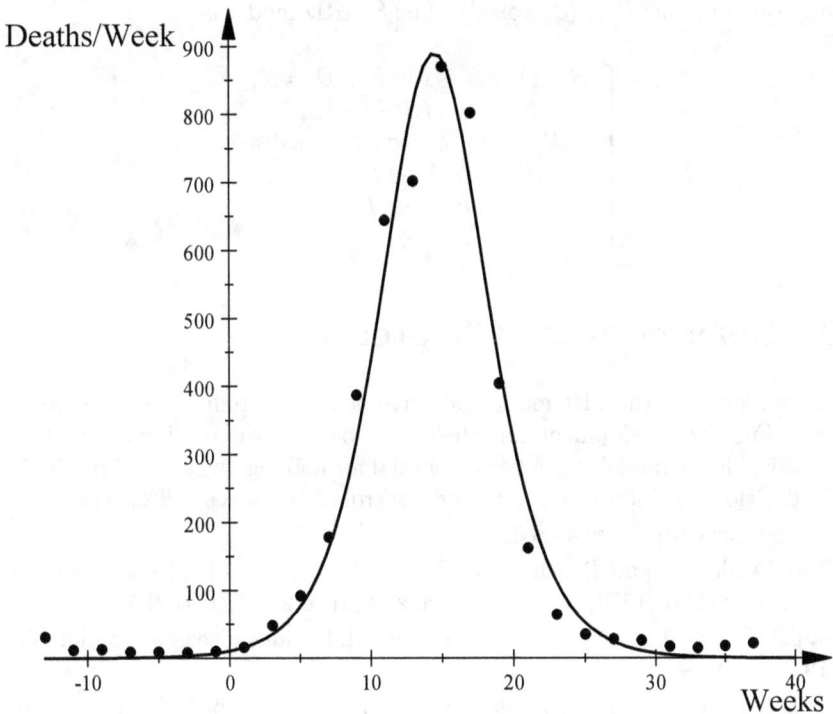

Figure 6.7: Bombay plague 1905-1906
data points & SIR model solution

reason including quarantine. The SEIR model is:

$$
\begin{cases}
S + E + I + R = N, \\[2mm]
S' = -bS\,(I + qE)\,/N, \\[2mm]
E' = +bS\,(I + qE)\,/N - \delta E \\[2mm]
I' = +\delta E - rI, \\[2mm]
R' = +rI.
\end{cases}
$$

Asymptomatic Carriers and SAIRD Models. Modeling accurately a disease like Covid that mutates rapidly has proven difficult. One aspect is that Covid includes an asymptomatic but infectious period. One proposed model of Olivera for the shorter term growth of a Covid epidemic includes groups $A(t)$

(asymptomatic) and $D(t)$ (deceased). The SAIRD model is

$$\begin{cases} S + A + E + I + R + D = N, \\ \quad S' = -b_I IS - b_A AS, \\ \quad A' = +b_I IS + b_A AS - aA \\ \quad I' = +aA - rI - \mu I, \\ \quad R' = +rI \\ \quad D' = +\mu I. \end{cases}$$

6.6   References for Chapter 6

We have seen that the SIR model can give accurate predictions of many epidemics. Further development has thus been based on extensions of SIR type models. Epidemic models have been essential for making "what if..." predictions to aid decisions on interventions in and control of the spread of disease.

References for Chapter 6.

R.M. Anderson and R. M. May, Population biology of infectious diseases, part 1, Nature 280 (1979) 361- 367, part 2, Nature 280 (1979) 455-461.

N.T.J. Bailey, The mathematical theory of infectious diseases, London, Griffin, 1975.

D. Defoe, A journal of the plague year, London, Oxford University Press, 1972.

K. Dietz, Epidemics and rumors: a survey, Journal of the Royal Statistical Society, 130 (1967), 505-528.

H.W. Hethcote, Qualitative analyses of communicable disease models, Math. Biosciences 28 (1976).

D. Kirschner, Using mathematics to understand HIV immune dynamics, Notices of the AMS, February, (1996) 191-202.

W.D. Kermack and A.G. McKendrick, A contribution to the mathematical theory of epidemics, Journal of the Royal Statistical Society, 115 (1927), 700-721.

P.E. Lekone and B.F. Finkenstadt, Statistical inference in a stochastic SEIR model with control interaction: Ebola as a case study, Biometrics 62(2006) 1170-1177.

R.M. May and R.M. Anderson, Transmission dynamics of HIV infection, Nature, 326 (1987), 137.

G. Oliveira, Refined Compartmental Models, Asymptomatic Carriers and COVID-19, ARXIV:2004.14780, 2020.

R. Ross, The prevention of malaria, second edition, London, Murray,.

H.E. Soper, Interpretation of periodicity in disease prevention, Journal of the Royal Statistical Society, 92 (1929), 34-73.

P. Waltman, Deterministic threshold models in the theory of epidemics, New York, Springer, 1974.

6.7 Exercises for Chapter 6

1. Compare the qualitative predictions of the parasite-host model to those of the *SIR* model.

2. Sketch either analytically or computationally the phase portrait of the *SIRS* model and compare its qualitative predictions to those of the *SIR* model.

3. Sketch the phase portraits of the SIR model for all three combinations of variables $S - I, R - S$ and $I - R$. Sketch the $I - R$ phase portrait.

4. Suppose an epidemic takes the path:

$$S \rightarrow E \rightarrow I \rightarrow R,$$

   where E are those exposed but not yet infected. Assume that a fixed percent of those exposed become infectious. Develop the resulting model.

5. State the Threshold Theorem for epidemics. Explain in words the main point of the theorem.

6. Rumors can spread in similar ways to epidemics. Develop a model for the spread of a rumor. Explain its assumptions and analyze its predictions. Compare the results with the rumor spread model from the exercises in Chapter 3.

7. Suppose vaccinations are administered at a constant rate $a > 0$.

   (a) Find the $I - S$ system arising from this additional assumption. What predictions can you make?

   (b) Show that the SIR epidemic ends after $t = N/a$.

8. Suppose the initial levels of $S(0) = S^*$ and $I(0) = I^*$ are known and consider the simplified, linear epidemic model.

$$
\begin{aligned}
S' &= -bIS^*, & S(0) &= S^*, \\
I' &= dS^*I - RI, & I(0) &= I^*, \\
R' &= RI. &
\end{aligned}
$$

   (a) Solve for S, I and R explicitly.

   (b) Find the threshold factor required to have an epidemic in this model.

9. Measles epidemics occur cyclically. By analogy to predator-prey oscillations, what sort of SIR model would lead to periodic orbits? Describe the assumptions behind your model and sketch its phase portrait.

10. Suppose the population is growing fast enough that a new supply of susceptibles are constantly being supplied. Generalize and describe the *SIR* and *SIRS* model to include this additional factor.

11. Consider the *SIRS* model. Eliminate R and find a 2×2 subsystem for (S, I). Sketch its phase portrait and describe how it differs from the *SIR* model.

12. Martin Braun proposed the following model of a gonorrhea epidemic:

$$
\begin{aligned}
x &= \quad \# \text{ infected females,} \\
y &= \quad \# \text{ infected males,} \\
x' &= \quad -ax + b(c - x)y, \\
y' &= \quad -dy + e(f - y)x.
\end{aligned}
$$

What are the assumptions leading to this model? Analyze the model.

13. Write down a dynamic model of the following: A rumor spreads among a population of $N + 1$ people starting with 1 person. At any time t there are three populations

$x(t) = $ those who haven't heard it
$y(t) = $ those actively spreading it
$z(t) = $ those who heard it and no longer care [thus don't spread it].
Assume that:

A1. The rate of spread is proportional to the number of contacts between those spreading it and those who haven't heard it.

A2. If two people meet who have already heard it, they then consider it old news and both stop spreading it [moving on to something else].

Chapter 7

Persistent Oscillations: Limit cycles

7.1 Oscillations in Nature

"Truth emerges more readily from error than from confusion." - Francis Bacon

Hudson bay company data of pelts of hare and lynx reveal a persistent 10 year population cycle. There is obviously a lot of noise and external forcing in this data and, perhaps as a consequence, the oscillations are far from cleanly periodic. In spite of these effects, the 10 year cycle period seems quite stable. The Lotka-Volterra equations were the first and ground breaking attempt to explain these cycles. These equations produce population cyclesof the sort plotted below in Figure 7.1.

The Lotka-Volterra model predicts cycles that develop by:

1. *The predator population is low so the prey population grows rapidly.*
2. *With more prey the predator population then grows rapidly.*
3. *The numerous predators suppress the prey population.*
4. *The predator population then decreases and the cycle repeats.*

This description seems plausible. However, it fails to fit three features of the hare-lynx data:

1. *The period of Lotka-Volterra solutions change every time a system perturbation nudges the solution off one cycle onto a nearby cycle. Thus, the solution period should be the least stable aspect of the system response!*

2. *The actual data of populations do not conform to the above picture: the lynx population cyclesactually leads the hare population cycle rather than following it as Lotka-Volterra suggests it should.*

3. *The predicted cyclesdisappear when the accuracy of the model is increased (as will be shown next in Section 7.1.1).*

153

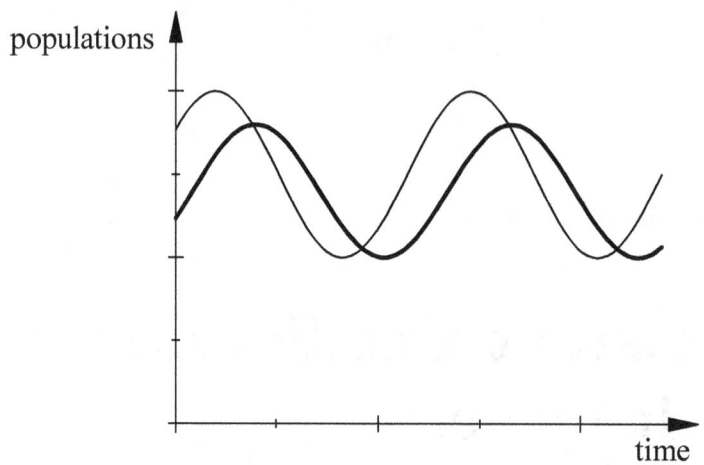

Figure 7.1: Prey population and **Predator population** vs time

This chapter builds the mathematical theory of limit cycles which are the mathematical key to resolving the first and third mysteries.

7.1.1   The logistic correction to the Lotka-Volterra model

"In theory there is no difference between theory and practice. In practice there is." - Yogi Berra

We shall see that the cyclespredicted by the Lotka-Volterra model are not persistent. A small change in the model that makes it more accurate causes all cyclesto disappear[1]. This will lead to a reconsideration of the mathematical architecture necessary to explain persistent oscillations in nature and the theory of limit cycles.

To begin, consider the Lotka-Volterra model with the small change that the prey is now assumed to undergo logistic growth in the absence of the predator:

$$x' = a\left(1 - \frac{x}{K}\right)x - bxy,$$
$$\text{and}$$
$$y' = -my + nxy.$$

The extra term, the model's logistic modification, is

$$\text{extra term} = -\frac{1}{K}ax^2 << 1$$

---

[1] This notion of persistence to model perturbations (rather than persisting over long time) is called structural stability.

when the carrying capacity K is large. Since this term is small and increases model accuracy its effect on the model solution is interesting. We prove the following.

Theorem. *For K large enough*

$$K > \frac{m}{n}$$

then the model has a unique positive equilibrium

$$(x^*, y^*) = \left(\frac{m}{n}, \frac{a}{b}(1 - \frac{m}{nK})\right).$$

If $a > b$ this equilibrium is a stable spiral.

The proof will use two properties of matrices: trace and determinant.

Definition. *Let A be the 2×2 matrix*

$$A = \begin{bmatrix} a & b \\ c & d \end{bmatrix}$$

then

$$\begin{aligned} trace(A) &= a + d, \\ \det(A) &= ad - bc. \end{aligned}$$

trace and determinant are related to eigenvalues as follows.

Theorem. *Let A be the 2×2 matrix*

$$A = \begin{bmatrix} a & b \\ c & d \end{bmatrix}$$

with eigenvalues λ_1, λ_2, then

$$\begin{aligned} trace(A) &= \lambda + \lambda_2, \\ \det(A) &= \lambda_1 \lambda_2. \end{aligned}$$

This is true more generally: the trace is the sum of all eigenvalues and the determinant is the product of all eigenvalues, Householder [H06]. For a 2×2 matrix the result is particularly useful since the two numbers trace and determinant completely determine the two eigenvalues.

Corollary. *Let A be the linearization matrix at an equilibrium (x^*, y^*). Then (x^*, y^*) is a stable spiral if*

$$trace(A) < 0 \quad \text{and} \quad \det(A) > 0.$$

Proof: *Then we have*

$$\lambda + \lambda_2 > 0 \text{ and } \lambda_1 \lambda_2 < 0..$$

Thus

$$\lambda_{1,2} = \alpha \pm \beta i$$

$(\alpha, \beta \ real)$ with $\alpha = \text{Re}(\lambda_{1,2}) < 0$.

We will use these results to analyze the Lotka-Volterra system with logistic growth leading to the proof of the above theorem.

Proof. Setting $x' = 0, y' = 0$ we have the nullclines:

$$x' = 0 \quad \Leftrightarrow \quad x = 0 \text{ or } \quad y = -\tfrac{a}{bK}x + \tfrac{a}{b}$$
$$\text{and}$$
$$y' = 0 \quad \Leftrightarrow \quad y = 0 \text{ or } \quad x = \tfrac{m}{n}.$$

The nullclines, $x = 0, y = 0, y = -\tfrac{a}{bK}x + \tfrac{a}{b}, x = \tfrac{m}{n}$, are plotted below in Figure 7.2. The analysis of this system closely parallels that of the Lotka-Volterra system

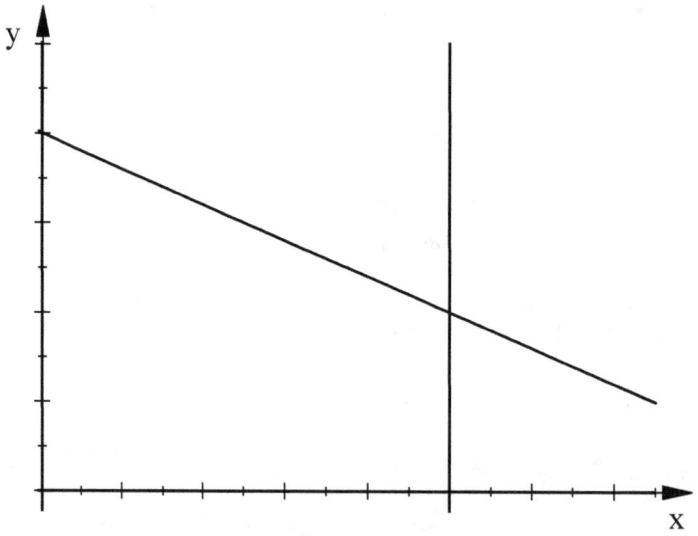

Figure 7.2:   Nullclines: $x = 0, y = 0, y = -\tfrac{a}{bK}x + \tfrac{a}{b}, x = \tfrac{m}{n}$ of the Lotka-Volterra model with logistic growth of prey.

so we shall proceed in larger steps. The trajectory directions are similar to those of the unmodified Lotka-Volterra system and suggest a counter clockwise rotation about the equilibrium (x^*, y^*). Indeed, the slope field is plotted next in Figure 7.3. Figure 7.4 presents a phase plane plot of the specific system $x' = (1 - x/5)x - xy, y' = -y + xy$. We will prove that this picture of a stable spiral is a typical phase plane for the logistic correction.

The unique positive equilibrium is easily found, as claimed, to be

$$(x^*, y^*) = \left(\frac{m}{n}, \frac{a}{b}(1 - \frac{m}{nK}) \right).$$

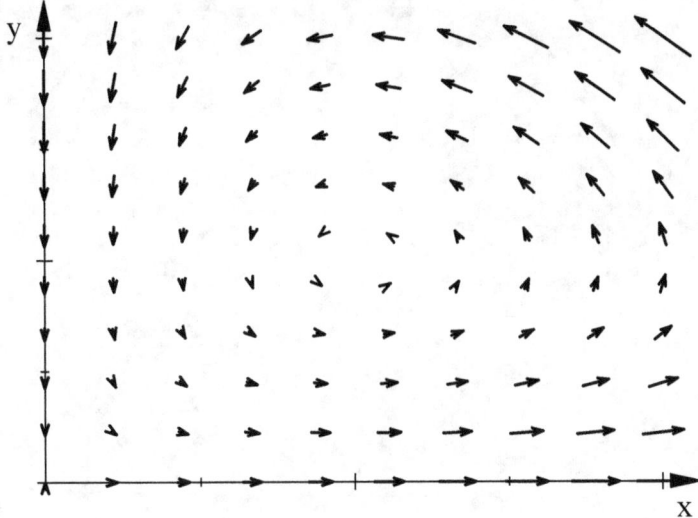

Figure 7.3:
Lotka-Volterra Model with logistic growth of prey
slope field indicates cycles about
equilibrium (x^*, y^*)

The linearization matrix at this equilibrium is also easily calculated to be

$$A = \begin{bmatrix} -\frac{2m}{nK}(a-b) & -\frac{bm}{N} \\ +\frac{a}{b}(n-\frac{m}{K}) & 0 \end{bmatrix}.$$

We then have

$$\det(A) = am\left(1 - \frac{m}{nK}\right) > 0 \text{ if } K > \frac{m}{n}$$

and

$$trace(A) = -\frac{2m}{nK}(a-b) < 0 \text{ if } a > b.$$

Thus the equilibrium is a stable spiral, completing the proof.

7.2 Introduction to Limit Cycles

"It's very easy to give simple examples that are not very interesting or interesting examples that are very difficult. If there isn't a simple, interesting case, forget it." - M. Atiyah

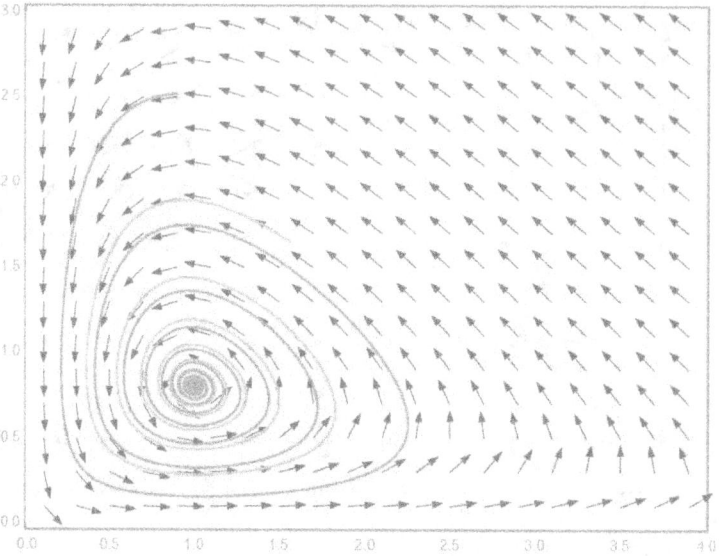

Figure 7.4: Phase plane of $x' = (1 - x/5)x - xy, y' = -y + xy$

The phase space of a two dimensional autonomous system

$$\begin{cases} x' & = & P(x,y), \\ y' & = & Q(x,y), \end{cases} \tag{7.1}$$

consists of combinations of nodes, non-intersecting trajectories, and cycles. cyclesare closed trajectories which are either separated from other closed trajectories (the case of limit cycles) or one of a family of adjacent closed trajectories (the case of a center).

Definition 44 *A closed trajectory C is a **limit cycle** if it is separated from all other closed trajectories. Precisely, C is a limit cycle if there is a tube about C of positive radius which contains no other closed trajectory.*

*A limit cycle C is a **stable limit cycle** if there is a tube of positive radius about C such that any trajectories which cross into that tube must approach C as t approaches infinity.*

*A limit cycle C is an **unstable limit cycle** if it is not stable. C is **semi stable** (and thus unstable) if trajectories entering on one side of C approach C but those on the other side of C in the tube do not*

The next two figures illustrate a center and a limit cycle.

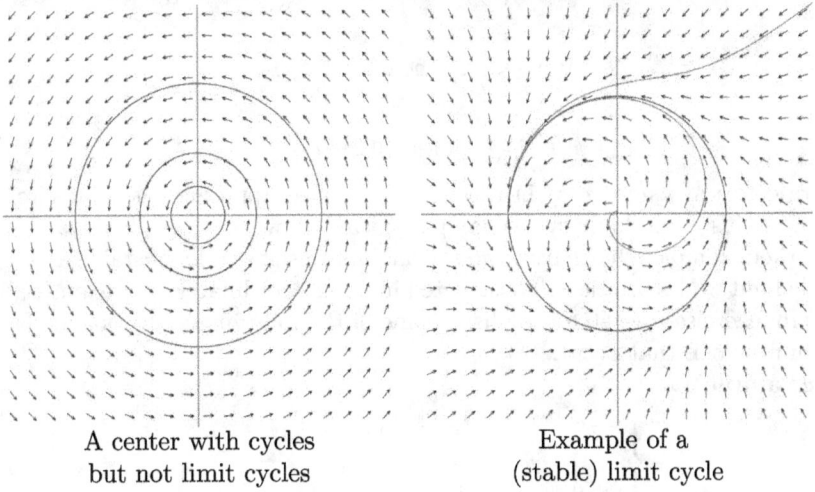

| A center with cycles but not limit cycles | Example of a (stable) limit cycle |

It is easy to construct examples of limit cycles by beginning in polar coordinates (r, θ). For example, consider the following system which has a limit cycle (the circle of radius one):

$$\begin{cases} x' & = & -y + x \left[1 - (x^2 + y^2)\right], \\ y' & = & +x + y \left[1 - (x^2 + y^2)\right]. \end{cases}$$

This example is constructed by beginning with the system below. The system is given in polar coordinates for which it is easy to see $r = 1$ is a limit cycle. After the desired limit cycle behavior is built into the polar system it is then transformed back to Cartesian coordinates using the chain rule. Indeed, consider the polar coordinate equations for $r(t)$ and $\theta(t)$:

$$\begin{cases} r' & = & r(1 - r^2) \\ \theta' & = & 1. \end{cases}$$

These equations have as one solution:

$$r(t) = 1, \quad \theta(t) = t,$$

which is a limit cycle (the circle of radius $1 : x^2 + y^2 = 1$) that winds counterclockwise around the origin (since $\theta(t)$ is increasing). We can check its stability by investigating whether $r(t)$ is increasing or decreasing when $r > 1$ and when $r < 1$. Indeed,

$$\begin{aligned} \text{if } r & > & 1, \text{ then } 1 - r < 0 \\ & & \text{and} \\ r' & = & r(1 + r)(1 - r) < 0. \end{aligned}$$

Thus, trajectories beginning *outside* ($r > 1$) the unit circle must approach the circle $r = 1$ as $t \to \infty$. If on the other hand, a trajectory begins inside the unit circle

$$\text{if } r \;\; < \;\; 1, \text{ then } 1 - r > 0$$
$$\text{and}$$
$$r' \;\; = \;\; r(1 + r)(1 - r) > 0,$$

so trajectories *inside* ($r < 1$) the unit circle also approach the unit circle as $t \to \infty$. This is the limit cycle example plotted above.

Other examples that have various combinations of stable, unstable and semi-stable limit cycles are easily constructed in polar coordinates. For example, it is a useful exercise to sketch the phase plane of the following systems and classify the limit cycles that occur:

Example 1:

$$\begin{cases} r' &= & r(r-1)(r-2)(r-3), \\ \\ \theta' &= & 1. \end{cases}$$

This has an equilibrium at $r = 0$ and three limit cycles, $r(t) = 1$, $r(t) = 2$ and $r(t) = 3$. We analyze its behavior between the cycles by checking signs as follows.

$$\begin{array}{llll} \text{If } 0 < r < 1, & r' = & +--- = - & \text{so } r(t) \text{ decreasing,} \\ \text{If } 1 < r < 2, & r' = & ++-- = + & \text{so } r(t) \text{ increasing,} \\ \text{If } 2 < r < 3, & r' = & +++- = - & \text{so } r(t) \text{ decreasing,} \\ \text{If } 3 < r < \infty, & r' = & ++++ = + & \text{so } r(t) \text{ increasing.} \end{array}$$

Example 2:

$$\begin{cases} r' &= & r(r-1)^2(r-2)^2(r-3)^2, \\ \theta' &= & 1. \end{cases}$$

This has an equilibrium at $r = 0$ and three limit cycles, $r(t) = 1$, $r(t) = 2$ and $r(t) = 3$. We analyze its behavior between the cycles by checking signs as before.

$$\text{If } 0 < r, \quad r' = \;\; ++++ = + \quad \text{so } r(t) \text{ is always increasing.}$$

Thus each limit cycle is semi-stable.

Example 3:

$$\begin{cases} r' &= & r(r-1)^2(r-2)(r-3)^2, \\ \theta' &= & 1. \end{cases}$$

This has an equilibrium at $r = 0$ and three limit cycles, $r(t) = 1$, $r(t) = 2$ and $r(t) = 3$. We analyze its behavior between the cycles by checking signs.

$$\begin{array}{llll} \text{If } 0 < r < 1, & r' = & ++-+ = - & \text{so } r(t) \text{ decreasing,} \\ \text{If } 1 < r < 2, & r' = & ++-+ = - & \text{so } r(t) \text{ decreasing,} \\ \text{If } 2 < r < 3, & r' = & ++++ = + & \text{so } r(t) \text{ increasing,} \\ \text{If } 3 < r < \infty, & r' = & ++++ = + & \text{so } r(t) \text{ increasing.} \end{array}$$

We leave it as an exercise to draw the phase plane and specify the stability of the limit cycles $r(t) = 1$, $r(t) = 2$ and $r(t) = 3$.

7.2.1 Constructing examples via polar coordinates

"Fundamentals, fundamentals. If you don't have them you'll run into someone else's." - *Virgil Hunter (Boxing trainer)*

The method of constructing examples by starting with a system in polar coordinates is useful so we review how to convert an $x - y$ system to and from polar. First suppose we have a system in polar

$$\theta' = A(r, \theta), \quad r' = B(r, \theta)$$

and we wish to find the equivalent $x - y$ system

$$x' = P(x, y), \quad y' = Q(x, y).$$

Starting with the fundamental change of variables

$$x = r \cos \theta, \quad y = r \sin \theta,$$

we regard all variables x, y, r, θ as functions of t. Differentiate with respect to t. The chain rule gives

$$x' = r' \cos \theta + r(-\sin \theta)\theta' \text{ and } y = r' \sin \theta + r(\cos \theta)\theta'$$

This can be written as matrix-vector multiplication with the 2×2 transformation matrix $\mathcal{A}$ as

$$\begin{bmatrix} x' \\ y' \end{bmatrix} = \mathcal{A}_{2\times 2} \begin{bmatrix} r' \\ \theta' \end{bmatrix}, \text{ where } \mathcal{A} := \begin{bmatrix} \cos \theta & -r \sin \theta \\ \sin \theta & r \cos \theta \end{bmatrix}_{2\times 2}.$$

The transformation matrix A is invertible (except at the origin) since

$$\det \mathcal{A} = r \cos^2 \theta + r \sin^2 \theta = r > 0.$$

There is a useful formula[2] for the inverse of a 2×2 matrix that says

$$\begin{bmatrix} a & b \\ c & d \end{bmatrix}^{-1} = \frac{1}{\det \mathcal{A}} \begin{bmatrix} d & -b \\ -c & a \end{bmatrix}$$

so that

$$\mathcal{A}^{-1} = \frac{1}{r} \begin{bmatrix} r \cos \theta & +r \sin \theta \\ -\sin \theta & \cos \theta \end{bmatrix}.$$

---

[2] This is easier to remember in words "*2 × 2 inverse is 1 over determinant times reverse the diagonal and switch signs of the off diagonal*".

$$[2 \times 2]^{-1} = \frac{1}{\text{determinant}} \begin{bmatrix} switch \searrow & negate \\ negate & \searrow switch \end{bmatrix}$$

It **only** works for the 2x2 case. The proof that it is correct is easy: multiply by A and verify that the product is I.

Case 1: Find the $x - y$ system, given the polar

$$\theta' = A(r, \theta), \quad r' = B(r, \theta).$$

Multiply through by $\mathcal{A}$. Using $\theta' = A(r, \theta), \quad r' = B(r, \theta)$ this gives

$$\begin{bmatrix} x' \\ y' \end{bmatrix} = \begin{bmatrix} \cos\theta & -r\sin\theta \\ \sin\theta & r\cos\theta \end{bmatrix} \begin{bmatrix} r' \\ \theta' \end{bmatrix} = \begin{bmatrix} \cos\theta & -r\sin\theta \\ \sin\theta & r\cos\theta \end{bmatrix} \begin{bmatrix} B(r, \theta) \\ A(r, \theta) \end{bmatrix}.$$

The system is reduced to $x - y$ form by replacing everywhere

$$r \Leftarrow \sqrt{x^2 + y^2}, r\cos\theta \Leftarrow x, r\sin\theta \Leftarrow y, \theta \Leftarrow \arctan(y/x).$$

Example of polar to Cartesian: Consider

$$\begin{cases} r' & = & r(1 - r^2) \\ \theta' & = & 1. \end{cases}$$

Then

$$\begin{aligned} \begin{bmatrix} x' \\ y' \end{bmatrix} &= \begin{bmatrix} \cos\theta & -r\sin\theta \\ \sin\theta & r\cos\theta \end{bmatrix} \begin{bmatrix} r' \\ \theta' \end{bmatrix} = \begin{bmatrix} \cos\theta & -r\sin\theta \\ \sin\theta & r\cos\theta \end{bmatrix} \begin{bmatrix} r(1 - r^2) \\ 1 \end{bmatrix} = \\ &= \begin{bmatrix} r(1 - r^2)\cos\theta - r\sin\theta \\ r(1 - r^2)\sin\theta + r\cos\theta \end{bmatrix}, \text{ using } x = r\cos\theta \text{ and } y = r\sin\theta \\ &= \begin{bmatrix} (1 - r^2)x - y \\ (1 - r^2)y + x \end{bmatrix}, \text{ using } r^2 = x^2 + y^2 \\ &= \begin{bmatrix} (1 - [x^2 + y^2])x - y \\ (1 - [x^2 + y^2])y + x \end{bmatrix}. \end{aligned}$$

Thus this is the system

$$\begin{cases} x' & = & (1 - [x^2 + y^2])x - y, \\ y' & = & (1 - [x^2 + y^2])y + x. \end{cases}$$

Case 2: Find the polar system, given the $x - y$ system

$$x' = P(x, y), \quad y' = Q(x, y).$$

Multiply the vector form by $\mathcal{A}^{-1}$. This gives

$$\begin{bmatrix} r' \\ \theta' \end{bmatrix} = \frac{1}{r} \begin{bmatrix} r\cos\theta & +r\sin\theta \\ -\sin\theta & \cos\theta \end{bmatrix} \begin{bmatrix} x' \\ y' \end{bmatrix}.$$

Thus, using $x' = P(x, y) = P(r\cos\theta, r\sin\theta)$, $y' = Q(x, y) = Q(r\cos\theta, r\sin\theta)$, the polar system is:

$$\begin{bmatrix} r' \\ \theta' \end{bmatrix} = \frac{1}{r} \begin{bmatrix} r\cos\theta & +r\sin\theta \\ -\sin\theta & \cos\theta \end{bmatrix} \begin{bmatrix} P(r\cos\theta, r\sin\theta) \\ Q(r\cos\theta, r\sin\theta) \end{bmatrix}.$$

Example of Cartesian to polar: Consider

$$\begin{cases} x' &= (1 - [x^2 + y^2])x - y, \\ y' &= (1 - [x^2 + y^2])y + x. \end{cases}$$

Then we have

$$\begin{bmatrix} r' \\ \theta' \end{bmatrix} = \frac{1}{r} \begin{bmatrix} r\cos\theta & +r\sin\theta \\ -\sin\theta & \cos\theta \end{bmatrix} \begin{bmatrix} x' \\ y' \end{bmatrix} = \frac{1}{r} \begin{bmatrix} r\cos\theta & +r\sin\theta \\ -\sin\theta & \cos\theta \end{bmatrix} \begin{bmatrix} (1 - [x^2 + y^2])x - y \\ (1 - [x^2 + y^2])y + x \end{bmatrix}.$$

The terms $1 - [x^2 + y^2]$ we can immediately replace by $1 - r^2$. This gives

$$\begin{aligned} \begin{bmatrix} r' \\ \theta' \end{bmatrix} &= \frac{1}{r} \begin{bmatrix} r\cos\theta & +r\sin\theta \\ -\sin\theta & \cos\theta \end{bmatrix} \begin{bmatrix} (1 - r^2)x - y \\ (1 - r^2)y + x \end{bmatrix} = \\ &= \frac{1}{r} \begin{bmatrix} r\cos\theta[(1 - r^2)x - y] + r\sin\theta[(1 - r^2)y + x] \\ -\sin\theta[(1 - r^2)x - y] + \cos\theta[(1 - r^2)y + x] \end{bmatrix} \end{aligned}$$

, using $x = r\cos\theta$ and $y = r\sin\theta$

$$= \frac{1}{r} \begin{bmatrix} r\cos\theta[(1 - r^2)r\cos\theta - r\sin\theta] + r\sin\theta[(1 - r^2)r\sin\theta + r\cos\theta] \\ -\sin\theta[(1 - r^2)r\cos\theta - r\sin\theta] + \cos\theta[(1 - r^2)r\sin\theta + r\cos\theta] \end{bmatrix}.$$

Expanding, using $\cos^2 + \sin^2 = 1$ and cancelling terms gives the polar system

$$\begin{cases} r' &= r(1 - r^2) \\ \theta' &= 1. \end{cases}$$

7.3 Poincaré-Bendixon Theory

"Les mathématiques sont comme le porc; tout en est bon." - *Lagrange*

The method of changing a system to polar coordinates is helpful only when the limit cycle is circular. This is not always true (in fact, seldom true in general). For example, the van der Pol equation

$$w'' - w'(1 - w^2) + w = 0, \tag{7.2}$$

becomes, with $x = w$, $y = w'$,

$$\begin{cases} x' &= y, \\ y' &= y(1 - x^2) - x, \end{cases}$$

and, in polar coordinates:

$$\begin{cases} r' &= r\sin 2\theta(1 - r^2\cos 2\theta), \\ \theta' &= -1 + \sin\theta(1 - r^2\cos 2\theta). \end{cases}$$

This polar form of the van der Pol equation does not really give any insight into possible limit cycles the van der Pol equation (7.2) might possess.

The existence of a stable limit cycles requires a tube that absorbs all trajectories entering the tube. However, this fact alone does not ensure that there is a limit cycles inside that tube. For example, the system

$$\begin{cases} r' & = & r(1-r), \\ \theta' & = & \sin\theta, \end{cases}$$

has no limit cycles. It has one saddle point and one node although it does possesses such an absorbing tube. Still, all trajectories entering a positively invariant tube must go somewhere. There is a beautiful theorem of Poincaré and Bendixon which states that the existence of an absorbing region is almost sufficient for the existence of a limit cycle in that region.

Definition 45 *A region D in the $x-y$ plane is **positively invariant** for the system (7.1) if, whenever a trajectory enters D at some time t^* it thereafter remains in D for all future times $t > t^*$.*

Theorem 46 *[The First Poincaré-Bendixon Theorem]. Suppose there is a bounded region D of the phase plane that is positively invariant for the system (7.1). Then, any trajectory entering D must either*

1. *approach an equilibrium point, or*

2. *spiral to a limit cycle of the system.*

We shall not prove this Poincaré-Bendixon theorem; see, for example, Hirsch and Smale [1974] for a proof. However, it gives immediately a simple sufficient criteria for the existence of a limit cycle.

Theorem 47 *[The Second Poincaré-Bendixon Theorem]. Suppose a closed and bounded region D is a positively invariant set for (7.1) and D contains no equilibriums. Then D must contain a limit cycle.*

Example: As an example, consider the system

$$\begin{cases} x' & = & y, \\ y' & = & -x + y(1 - 3x^2 - 2y^2), \end{cases} \qquad (7.3)$$

which comes from the second order equation

$$w'' - (1 - 3w^2 - 2\left(w'\right)^2)w' + w = 0.$$

Clearly, $(0,0)$ is an equilibrium. We can get an equation for r' directly from the $x-y$ form (and without going through the steps to change variables given above) as follows[3]. Since $r^2 = x^2 + y^2$ differentiation gives

$$\frac{d}{dt}r^2 = 2rr' = 2xx' + 2yy'.$$

---

[3]This is another useful trick/formula.

For the above equation this gives

$$
\begin{aligned}
rr' &= x\,(y) + y\left(-x + y(1 - 3x^2 - 2y^2)\right)\\
&= y^2(1 - 2x^2 - 2y^2 - x^2)\\
&= (r\sin\theta)^2\left(1 - 2r^2 - (r\cos\theta)^2\right).
\end{aligned}
$$

Thus,

$$
r' = r\sin^2\theta\left(1 - 2r^2 - (r\cos\theta)^2\right).
$$

Now consider the sign of r' on two circles $r = 1/4$ and $r = 1$:

$$
\text{when } r = 1/4: \quad r' = \tfrac{1}{4}\sin^2\theta\left(\tfrac{14}{16} - \tfrac{1}{16}(\cos\theta)^2\right) \quad \geq 0
$$
$$
\text{and}
$$
$$
\text{when } r = 1: \quad r' = -\sin^2\theta\left(1 + (\cos\theta)^2\right) \quad \leq 0
$$

Thus, the annulus D given by:

$$
D := \{(x,y)\mid 1/16 \leq x^2 + y^2 \leq 1\}
$$

is a closed and bounded positively invariant set. It is easy to check that the only fixed point for equation (7.3) is $(0,0)$. In particular equation (7.3) has no equilibrium points inside the annulus D. Thus, equation (7.3) has a limit cycle inside D.

The second Poincaré-Bendixon theorem gives sufficient conditions for the existence of a limit cycle. The third Poincaré-Bendixon theorem gives a sufficient condition for non existence of a limit cycle in a region.

Theorem 48 *[Third Poincaré-Bendixon Theorem]. Consider the system (7.1). If D is a simply connected region (that is, it contains no holes) such that*

$$
P_x(x,y) + Q_y(x,y)
$$

has one sign in D, then there can be no limit cycles of (7.1) entirely inside of D.

Example: For example, applying the third Poincaré-Bendixon theorem to the above system (7.3) we calculate:

$$
P_x(x,y) + Q_y(x,y) = 1 - 3x^2 - 6y^2.
$$

Thus, the system (7.3) can have no limit cycles entirely inside the ellipse

$$
\{(x,y)\mid 3x^2 + 6y^2 \leq 1\}.
$$

Example: Consider the generalized Lotka-Volterra model

$$
\begin{aligned}
x' &= (a_0 + a_1 x + a_2 y)x,\\
&\text{and}\\
y' &= (b_0 + b_1 x + b_2 y)y.
\end{aligned}
$$

Applying the third Poincaré-Bendixon theorem gives the following theorem.

Theorem. *If the coefficients satisfy*

$$a_0 + b_0 > 0,$$
$$2a_1 + b_1 > 0,$$
$$a_2 + 2b_2 > 0,$$

the generalized Lotka–Volterra model has no periodic solutions in the positive quadrant.

proof: *We calculate:*

$$P_x(x,y) + Q_y(x,y) = (a_0 + b_0) + (2a_1 + b_1)x + (a_2 + 2b_2)y$$

which is positive under the stated conditions.

7.4 Birth of Oscillations: Hopf bifurcations

"Things work in cycles." - Joan Jett

Frequently mathematical models involve various parameters, such as birth rates, carrying capacities and so on, and the behavior of the model changes as a parameter varies. The Hopf bifurcation theorem[4] describes a situation when an equilibrium point suddenly changes into a periodic motion as a parameter is slowly varied. For example, consider the ordinary differential equations

$$\begin{cases} x' & = & P(x,y;\mu), \\ \\ y' & = & Q(x,y;\mu), \end{cases} \tag{7.4}$$

where the parameter μ has been explicitly inserted. Suppose the system (7.4) has an equilibrium $(x^*(\mu), y^*(\mu))$, which we assume depends upon μ continuously. In certain cases, $(x^*(\mu), y^*(\mu))$ can change from a stable equilibrium as μ increases past some threshold into an unstable equilibrium surrounded by a stable limit cycle. In an experiment, unstable states are just not observed. Thus, experimentally, this mathematical behavior is manifested in the experiment suddenly jumping from a rest state to an oscillatory motion - dramatic indeed.

To illustrate this let us first consider an example of an ordinary differential equation which is amenable to exact analysis.

Example. For μ a real parameter consider the system:

$$\begin{cases} x' & = & \mu x - y - x(x^2 + y^2), \\ \\ y' & = & x + \mu y - y(x^2 + y^2). \end{cases} \tag{7.5}$$

---

[4]Due to Eberhard Hopf. Sometimes it is called the Poincare-Andonov-Hopf bifurcation theorem.

To analyze (7.5), first $(0,0)$ is a critical point of the system for all real values of μ. The linearized system[5] about $(0,0)$ is given by

$$
\begin{array}{rcl}
u' & = & [\mu - (x^2 + y^2) - x(2x)]u + [-1 - 2xy]v \\
v' & = & [1 - 2xy]u + [\mu - (x^2 + y^2) - y(2y)]v,
\end{array}
\Bigg|_{(x,y)=(0,0)}
$$

with both equations evaluated at $x = 0$ and $y = 0$ or, after simplification,

$$
\begin{cases}
u' & = \mu u - v \\
v' & = u + \mu v.
\end{cases}
\tag{7.6}
$$

The linearization matrix is

$$
A = \begin{bmatrix} \mu & -1 \\ +1 & \mu \end{bmatrix} \text{ with eigenvalues: } \lambda = \mu \pm i.
$$

For $\mu < 0$, the equilibrium $(0,0)$ of the linearized system (7.6) is a stable spiral and hence for (7.5) as well. For $\mu > 0$, the equilibrium $(0,0)$ of (7.6) is an unstable spiral and (7.5) is thus also. For $\mu = 0$, the equilibrium $(0,0)$ for the linearized problem (7.6) is a center. Thus, the equilibrium point $(0,0)$ for (7.5) could be a center or a spiral. We shall investigate the nonlinear problem (7.5) at $\mu = 0$ by switching to polar coordinates.

Indeed, transforming (7.5) to polar coordinates and simplifying algebraically gives:

$$
\begin{cases}
r' & = r(\mu - r^2) \\
\theta' & = 1.
\end{cases}
\tag{7.7}
$$

For $\mu = 0$ the non-trivial solution to (7.7) is

$$
\theta(t) = t + C_1 \quad \text{and} \quad r(t) = \frac{1}{\sqrt{2t + C_2}}.
$$

As $r(0) > 0$, we have $C_2 > 0$. Hence $r(t) \to 0$ as $t \to \infty$. As a result, the equilibrium $(0,0)$ of the system (7.5) is a stable spiral for $\mu = 0$ as well.

The interesting behavior of (7.5) occurs around (x^*, y^*) as μ crosses through zero. Consider equation (7.7) for $\mu > 0$. We have as a solution

$$
r(t) = \sqrt{\mu} \text{ and } \theta(t) = t + C.
$$

For $0 < r < \sqrt{\mu}$ we have $r' > 0$ and for $r > \sqrt{\mu}$, we have $r' < 0$. Thus, the circular trajectory $r(t) = \sqrt{\mu}$ is a stable limit cycle which spontaneously appears for $\mu > 0$. The size of the stable limit cycle increases as μ increases and the critical point $(x^*(\mu), y^*(\mu))$ changes from a stable spiral to an unstable spiral as μ crosses through the critical value 0. This is precisely the picture predicted by the celebrated Hopf bifurcation theorem.

Theorem 49 *(Hopf bifurcation Theorem) Suppose (7.4) has a critical point at $(0,0)$ for all μ, and that $(0,0)$ is asymptotically stable when $\mu = \mu^*$. Suppose that the eigenvalues $\lambda_1(\mu)$ and $\lambda_2(\mu)$ of the linearized system satisfy:*

---

[5] This system is written so the linearization at $(0,0)$ is just the linear terms on the RHS.

(a.) $\lambda_1(\mu)$ and $\lambda_2(\mu)$ are purely imaginary when $\mu = \mu^*$:

$$Re(\lambda_j(\mu^*)) = 0, \text{ for } j = 1, 2.$$

(b.) $\lambda_j(\mu)$ satisfy

$$\frac{d}{d\mu}\left[Re(\lambda_j(\mu))\right]|_{\mu=\mu^*} > 0 \text{ , for } j = 1, 2.$$

Then, for some $\epsilon > 0$:

1. for $\mu^* - \epsilon < \mu < \mu^*$, the critical point $(0,0)$ is a stable spiral and

2. for $\mu^* < \mu < \mu^*+\epsilon$, the critical point $(0,0)$ is an unstable spiral surrounded by a stable limit cycle whose size increases with μ.

The value $\mu = \mu^*$ has special interest.

Definition 50 *The value $\mu = \mu^*$ in the Hopf bifurcation theorem is called a **bifurcation point** for (7.4).*

The Hopf bifurcation theorem can be extended slightly to include the case of $(x^*(\mu^*), y^*(\mu^*))$ is a center rather than a stable spiral. In this case the picture, described by a degenerate Hopf bifurcation, is not as complete as with a full Hopf bifurcation.

Theorem 51 (Hopf bifurcation Theorem, $(x^*(\mu^*), y^*(\mu^*))$ a center) *Suppose $(x^*(\mu), y^*(\mu))$ is asymptotically stable for $\mu < \mu^*$ and unstable for $\mu > \mu^*$. Suppose that the eigenvalues, $\lambda_j(\mu), j = 1, 2$ of the linearization at the equilibrium point satisfy*

(a.) $Im[\lambda_j(\mu^*)]$ is non-zero, and

(b.) $\frac{d}{d\mu}[Re(\lambda_j(\mu))]|_{\mu=\mu^*} > 0.$

Then, for some $\varepsilon > 0$ and for some μ with $|\mu - \mu^*| < \varepsilon$ a closed orbit exists around (x^*, y^*).

The picture of the local behavior is thus as follows.

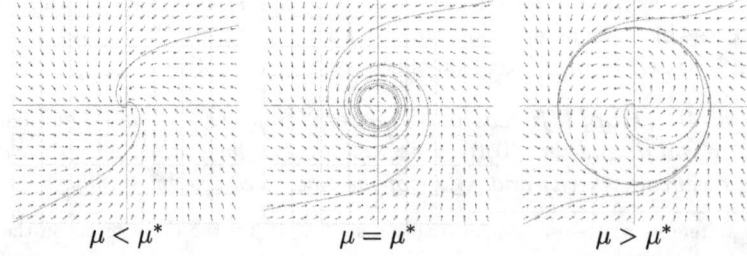

$\mu < \mu^*$ $\mu = \mu^*$ $\mu > \mu^*$

Example. Consider the system

$$
\begin{cases}
x' &= \mu x - 2y - 2x[x^2 + y^2]^2, \\
y' &= 2x + \mu y - 2y[x^2 + y^2]^2.
\end{cases}
\tag{7.8}
$$

We shall show that (7.8) undergoes a Hopf bifurcation when $\mu = 0$. The point $(0,0)$ is indeed a critical point of (7.8) for all μ. As in the previous example, the linearization of (7.8) is:

$$
\begin{cases}
u' &= \mu u - 2v \\
v' &= 2u + \mu v.
\end{cases}
$$

The eigenvalues of the 2×2 linearization matrix:

$$
\begin{bmatrix} \mu & -2 \\ 2 & \mu \end{bmatrix}
$$

are easily found[6] to be

$$
\lambda_{1,2}(\mu) = \mu \pm 2i.
$$

Thus, parts (a) and (b) of the theorem are satisfied. That $(0,0)$ is asymptotically stable for $\mu = 0$ can be verified by converting (7.8) to polar coordinates or by using the idea off a Lyapunov function introduced in Chapter 1.

Definition 52 *A function $V(x,y)$ is called a Lyapunov function if it has the three properties that: $V(x,y) \geq 0$; $V = 0$ if and only if $x(t) = x^*$ and $y(t) = y^*$ and $\frac{d}{dt}V(t) < 0$ along non-equilibrium solutions $x(t), y(t)$ of the system.*

The condition that $\frac{d}{dt}V(t) < 0$ can be weakened. Naturally all the conditions need only in some region around the equilibrium to conclude the equilibrium is stable.

For the $\mu = 0$ system

$$
\begin{cases}
x' &= -2y - 2x[x^2 + y^2]^2, \\
y' &= 2x - 2y[x^2 + y^2]^2,
\end{cases}
$$

the function

$$
V(x,y) = x^2 + y^2
$$

---

[6] *Another approach to finding the eigenvalues:* Note that

$$
\begin{bmatrix} \mu & -2 \\ 2 & \mu \end{bmatrix} = 2 \begin{bmatrix} 0 & -1 \\ +1 & 0 \end{bmatrix} + \mu \begin{bmatrix} 1 & 0 \\ 0 & 1 \end{bmatrix}.
$$

This is a function $f(z) = 2z^1 + \mu z^0$ of the matrix

$$
\begin{bmatrix} 0 & -1 \\ +1 & 0 \end{bmatrix}
$$

which has eigenvalues $+i$, $-i$. The spectral mapping theorem implies the eigenvalues of the function of this matrix are $f(z) = 2(\pm i) + \mu$.

can be verified to satisfy the above three criteria for a Lyapunov function as follows. For the $\mu = 0$ system, multiply the first equation of (7.8) by x, the second of (7.8) by y and add. Using $xx' + yy' = \frac{1}{2}\frac{d}{dt}[x^2 + y^2]$ gives

$$
\begin{aligned}
\frac{1}{2}\frac{d}{dt}[x^2 + y^2] &= -2xy - 2x^2[x^2 + y^2]^2 + 2xy - 2y^2[x^2 + y^2]^2 \\
&= -2[x^2 + y^2]^3 < 0 \text{ for } (x, y) \neq (0, 0).
\end{aligned}
$$

Thus, $V(x, y) = x^2 + y^2$ is strictly decreasing and $(0, 0)$ is asymptotically stable for $\mu = 0$. From the Hopf bifurcation theorem we can now conclude that (7.8) undergoes a Hopf bifurcation at $\mu^* = 0$ and has a stable limit cycle for $\mu > 0$.

Example. Consider the second order problem

$$\theta'' + (\theta')^3 - \mu\theta' + \theta = 0.$$

Let $x = \theta, y = \theta'$:

$$
\begin{aligned}
x' &= y, \\
y' &= -y^3 + \mu y - x.
\end{aligned}
$$

This is constructed so that $(0, 0)$ is an equilibrium(and thus the linearization at $(0, 0)$ is easy). The linearization at $(0, 0)$ is

$$
\frac{d}{dt}\begin{bmatrix} u \\ v \end{bmatrix} = \begin{bmatrix} 0 & +1 \\ -1 & \mu \end{bmatrix} \begin{bmatrix} u \\ v \end{bmatrix}.
$$

The eigenvalues of the linearization matrix are easily calculated:

$$
0 = \det\left\{ \begin{bmatrix} 0 & +1 \\ -1 & \mu \end{bmatrix} - \lambda \begin{bmatrix} 1 & 0 \\ 0 & 1 \end{bmatrix} \right\}
$$

or

$$
-\lambda(\mu - \lambda) + 1 = 0.
$$

The quadratic formula gives

$$
\lambda_{1,2} = +\frac{\mu}{2} \pm \left(\frac{1}{2}\sqrt{4 - \mu^2} \right) i.
$$

Thus for $-2 < \mu < +2$ the eigenvalues are complex and $\text{Re}(\lambda) = +\frac{\mu}{2}$. This changes sign from negative to positive as μ crosses from negative to positive through $\mu = 0$. Thus all the requirements for a full Hopf bifurcation at $\mu = 0$ are verified except asymptotic stability when $\mu = 0$. To check this set $\mu = 0$. We calculate the Lyapunov function as follows

$$
\begin{array}{llll}
(x' = y) & \times x & \Rightarrow & x'x = xy \\
(y' = -y^3 - x) & \times y & \Rightarrow & y'y = -xy - y^4 \\
\text{and} & \text{add} & \Rightarrow & \frac{1}{2}\frac{d}{dt}(x^2 + y^2) = -y^4 < 0 \text{ for } y \neq 0.
\end{array}
$$

Thus, the system a $\mu = 0$ is asymptotically stable and a Hopf bifurcation occurs at $\mu = 0$.

7.5 References for Chapter 7

A.A. Andronov and C.E. Chaikin, Theory of Oscillations, Princeton University Press, Princeton, 1949.

M.W. Hirsch and S. Smale, Differential equations, dynamical systems and linear algebra, Academic Press, London, 1974.

A. Householder, The theory of matrices in numerical analysis, Dover, 2006.

J.E. Marsden and M. McCracken, The Hopf bifurcation and its applications, Applied Mathematical Sciences, Volume 19, Springer Verlag, New York, 1976.

7.6 Exercises for Chapter 7

1. Consider the predator-prey model with logistic corrections.

$$\begin{cases} x' &= x(1-y-ax) \\ y' &= -y(1-x-by) \end{cases}$$

   (a) What assumptions lead to this model?

   (b) Find the non zero critical point of this model and show that it is a center if a and b are both 0. For what other values of a and b is the linearized problem a center?

   (c) Show that the equilibrium point is a stable spiral if $b=0$ and $a>0$. Sketch a representative phase portrait. What occurs for more general values of a and b? What does this say about the of the Lotka-Volterra model to the phenomena it describes?

2. Sketch a phase portrait consistent with (a) One unstable limit cycle, three equilibria: one saddle and two unstable nodes. (b) one stable focus and two limit cycles: one stable and one unstable.

3. a. Show that periodic solutions of the normal Lotka-Volterra model must cross the line

$$y = \frac{n}{b}x + \frac{a-m}{b}.$$

   b. Show that the following has no periodic solutions in the positive quadrant:

$$\begin{aligned} x' &= axy, \\ y' &= bxy. \end{aligned}$$

4. Consider the ordinary differential equation where all constants are positive.

$$\begin{cases} x' &= \frac{a}{A+ky} - b, \\ y' &= cx - d. \end{cases}$$

(a) Sketch the phase portrait of this system for several different combinations of parameters.

(b) For what combinations of parameters do you expect or observe oscillations in the solutions of this system?

5. Consider the second order equation which follows, written as a first order system.

$$w'' + w'^3 - \mu w' + w = 0.$$

(a) Rewrite this as an equivalent first order system. Find its linearization about the origin. Show that it undergoes a Hopf bifurcation at $\mu = 0$.

(b) Sketch its phase portrait for $\mu = -0.5, 0, 0.5$.

6. For the systems below, find and classify all limit cycles: $\theta' = 1$ and

a. $r' = r(r - 1)(r - 2)$

b. $r' = r^2(r - 1)(r - 2)$,

c. $r' = r(r - 1)^2(2 - r)(3 - r)$,

d. $r' = r(r - 1)(r - 2)(r - 3)$,

e. $r' = r(r - 1)^2(r - 2)^2(r - 3)^2$

f. $r' = r(r - 1)^2(r - 2)(r - 3)^2$.

7. [A system that is simple in polar coordinates.] Consider the system below.

(a) Show that $(0, 0)$ is a center for the linearized problem.

(b) Convert it to polar coordinates and show that $(0, 0)$ is a spiral point for the nonlinear system by showing $r(t) \to 0$ as $t \to \infty$.

$$x' = -y - x\sqrt{x^2 + y^2},$$
$$y' = x - y\sqrt{x^2 + y^2}.$$

(c) A simpler way to show this: Try multiplying the equations by ax and by, adding and then picking a, b to make the sum a simple equation for $ax^2 + by^2$.

8. State (a) the Hopf bifurcation theorem, and (b) the Poincaré - Bendixon theorems.

9. (a) Show that the equation below can have no periodic solutions lying entirely in a region where $f(\cdot)$ has one sign:

$$w'' + f(w)w' + g(w)w = 0.$$

(b) Show that the system below has no periodic solutions

$$x' = y ,$$
$$y' = 1 + x^2 - (1 - x)^2 y.$$

10. Sketch phase portraits consistent with

    (a) one unstable limit cycle, one saddle point and two unstable nodes.

    (b) one stable focus, one stable limit cycle and one unstable limit cycle.

11. Show that the system

$$\begin{aligned} x' &= 1 - xy, \\ y' &= x. \end{aligned}$$

comes from a second order ODE written as a first order system, and can have no limit cycles contained entirely in the first quadrant.

12. Consider the van der Pol equation (7.2). Find a positive invariant region containing a limit cycle.

13. Consider

$$w'' + (w^2 - \mu)w' + w + w^4 w' = 0.$$

    (a) Write it as a first order system in the standard way $(x = w, y = w')$. Find the linearization about the critical point $(0,0)$.

    (b) Show that the system has a bifurcation point μ_0.

    (c) Show $(0,0)$ is asymptotically stable for $\mu = \mu_0$.

    (d) Sketch the phase portrait near $(0,0)$.

14. Consider the system

$$\begin{aligned} x' &= y - x^3, \\ y' &= -x + \mu y - x^2 y. \end{aligned}$$

which has an equilibrium at $(0,0)$.

    (a) Find the eigenvalues of the linearization at $(0,0)$.

    (b) Using the Lyapunov function $V(x,y) = x^2 + y^2$ show that the system is asymptotically stable for $\mu \leq 0$.

    (c) Show that it undergoes a Hopf bifurcation at $\mu = 0$.

15. Suppose that a system has exactly one equilibrium point, n limit cycles and no other periodic orbits.

    (a) Explain why a stable limit cycle must be adjacent to an unstable one but an unstable one may be next to either.

    (b) Let c_n be the number of possible stability configurations of n nested limit cycles. Thus [drawing pictures we find] $c_1 = 2$, $c_2 = 3$, $c_3 = 5$. Show that this sequence is the Fibonacci sequence.

16. Consider the system in polar coordinates given below. Sketch its phase portrait and label all significant structures such as limit cycles indicating their stability.

$$\theta' = 1,$$
$$r' = r^2(r-1)(2-r).$$

17. Consider the system below which has an equilibrium at $(0,0)$:

$$x' = y - x^5,$$
$$y' = -x - x^2 y$$

Prove that $(0,0)$ is asymptotically stable.

18. For the system below, find a first integral. Using it, Show that the trajectories are ellipses in the phase plane. Using the first integral, find a quadratic or higher order term to add to the equation that converts $(0,0)$ to a stable spiral.

$$x' = b^2 y,$$
$$y' = -a^2 x.$$

Chapter 8

Oscillations in the Holling-Tanner Model

This chapter develops models that more accurately capture predator-prey interactions and predict the existence of persistent population cycles.

8.1 The Development of Predator-Prey Models

"... the current state of the art in population dynamics is that at present we can make quantitative testable models only of laboratory systems which are themselves only a caricature of reality." - p. 2 in: Nisbet and Gurney, Modeling fluctuating populations.

Lotka in 1925 and Volterra in 1927 independently developed the first predator-prey model studied in Chapter 4

$$x' = ax - bxy \quad \text{and} \quad y' = -my + nxy.$$

The model's solution produced periodic solutions, the simplest possible cycles, and explained a (qualitative) harvesting effect that had not been understood previously. However, many attempts to use it to forecast numbers of actual population failed. The model did not produce stable limit cycles as solutions and failed in other ways as well. There have been many models proposed for interacting species, most, including the Holling-Tanner model developed in this chapter, attempting to capture the essential behavior of cycles with stable periods by incorporating some aspect of predator prey interactions omitted by the Lotka-Volterra model. The most obvious omission from Lotka-Volterra is that the assumption the prey undergoes exponential growth and not logistic growth in the absence of the predator. Unfortunately, this correction by itself moves the system behavior further from what is observed.

Motivated by this fact that simple changes to the Lotka-Volterra model make the model predictions change dramatically, many extensions of the Lotka-Volterra model have been studied. In 1934 Gause (who wrote the famous paper

Do Snowshoe hare eat lynx?) proposed the model

$$x' = ax - p(x)y \quad \text{and} \quad y' = -my + np(x)y.$$

Here $p(x)$ and n denote

$$p(x) \quad = \quad \text{capture rate of prey by predators,}$$
$$n \quad = \quad \text{conversion rate of prey into predators,}$$

and $p(x)$ satisfies $p(0) = 0$, $p'(x) > 0$. The new parameters $p(x)$ and n were intended to be closer to observables. A more complex form of $p(x)$ can also incorporate more complex situations.

A next step occurred in a 1963 paper of Rosenzweig and McArthur. They studied a model including logistic prey growth and a predation rate with satiation. The predation rate

$$p(x) = \frac{bx}{D + x}$$

had been suggested by Holling in 1959 based on experimental data[1] and incorporated a satiation effect. The Holling-Tanner model was a further development of these ideas. Holling suggested three such predation rates. The one above is Holling type II. Holling type III (not analyzed herein) is

$$p(x) = \frac{bx^2}{D + x^2}.$$

8.2   The Holling-Tanner Predator-Prey Model

"A lion's work hours are only when he's hungry; once he's satisfied, the predator and prey live peacefully together." - Chuck Jones, Season 5, Episode 19, 'Rite of Passage' - spoken by Emily Prentiss.

Periodic orbits observed in nature must be stable to higher order perturbations; thus simplified models attempting to describe these cycles must exhibit stable limit cycles. The Holling-Tanner model is one successful attempt to find a model of predator-prey interactions which possess these types of stable limit cycles. The Holling-Tanner model is based on the following three assumptions.

Assumption A1. *Competition between prey for limited resources leads to logistic growth of the prey population in the absence of predators. The carrying capacity is large with respect to other system parameters.*

Assumption A2. *The prey's population is reduced by predation. The amount of predation is given by a predation rate times the number of predators. The predation rate depends on the prey population levels in such a way*

---

[1] It is widely mis-reported that the entomologist C.S. Holling based his suggested predation rate on field data. In fact it was derived from an experiment with students and sanding disks. He put blindfolded students (predators) around a table with a random distribution of sandpaper disks (prey). The students put their fingers on the table at random and he counted the number of finger-disk contacts as a function of disk density and student numbers.

as to increase to an upper limit, w, when the predator's appetite is satisfied (meaning, in the case of a relatively large number of prey).

Assumption A3. *The predator's population level obeys a* logistic *growth equation where the carrying capacity depends on the number of available prey per predator. Specifically, assume that it takes J prey per unit time to support one predator.*

Assumption A1 thus describes logistic growth of the prey with carrying capacity, K, fixed by the prey's environment. Assumption A2 is a predation assumption with the relative predation rate declining (i.e., the predation rate approaching a finite upper limit) as the predators become satiated, certainly a reasonable condition, Figure 8.2. Assumption A3 can be restated to assert that the carrying capacity of predators is x/J where x is the prey's population level. Thus, the predator's growth rate in the logistic model is $s(1 - Jy/x)$, depicted in Figure 8.1.

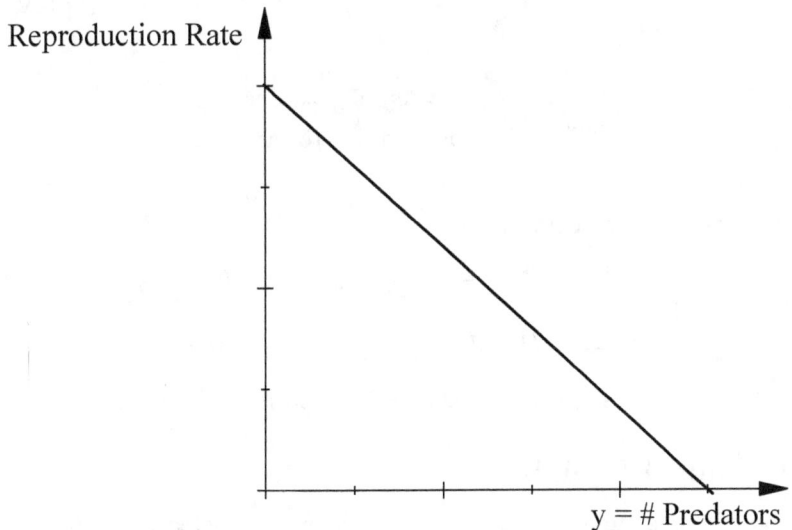

Figure 8.1:   Logistic carrying capacity = #prey / J
so predator reproduction rate = s(1-Jy/x)

The system of ordinary differential equations derived from these three assumptions is called the Holling-Tanner model. Let

$$x(t) = \text{population level of prey,}$$
$$\text{and}$$
$$y(t) = \text{population level of predator.}$$

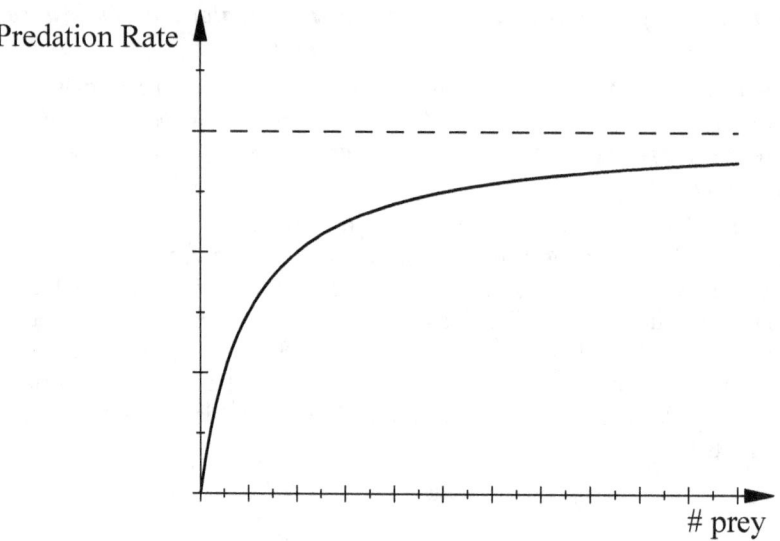

Figure 8.2: Predation rate $\frac{wx}{D+x}$ bounded due to satiation

The Holling-Tanner model is as follows:

$$\begin{cases} x' & = & r\left(1-\frac{x}{K}\right)x - \frac{wx}{D+x}y, & x(0) = x_0, \\ y' & = & s\left(1 - J\frac{y}{x}\right)y, & y(0) = y_0, \end{cases} \tag{8.1}$$

where the parameters r, s, w, K, D and J are positive constants.

8.2.1   Model Equilibria

Since populations are always non negative we are interested in the behavior of solutions to the system (8.1) in the first quadrant. There are two critical points of (8.1) lying in the first quadrant. The first is $(0,0)$. The non-trivial critical point (x^*, y^*) is found when $y \neq 0$ in the second equation of (8.1). This implies that

$$1 - \frac{Jy}{x} = 0.$$

This gives

$$x^* = Jy^* \tag{8.2}$$

Inserting this into the first equation of (8.1), the equation can be manipulated to be quadratic in y. We then obtain y^* from the quadratic formula

$$y^* = \frac{rJK - DrJ - wK + \sqrt{(wK + DrJ - rJK)^2 + 4r^2J^2DK}}{2rJ^2} \tag{8.3}$$

To understand the main features of the Holling-Tanner model we need to first recall some of the basic ideas of limit cycles. First recall the definition of a limit cycle.

Definition 53 *A limit cycle C is a closed trajectory in the phase plane that is isolated from any other closed trajectory. Specifically, there is a tube about C containing no other closed trajectories.*

Recall that all limit cycles are divided into three (overlapping) types:
Stable limit cycles, *in which trajectories spiral into C from both sides as $t \to \infty$,*
Unstable limit cycles, *in which trajectories spiral away from C as $t \to \infty$, from at least one side, and*
Semi-stable limit cycles*(a sub-type of unstable limit cycle), in which, as $t \to \infty$, trajectories spiral to C from one side and away from C on the other side of C.*

Poincaré-Bendixon Theory gives a sufficient condition for the existence of a limit cycle in a region: If a positive invariant region does not contain a stable equilibrium, then it must contain a stable limit cycle.

8.3 Analysis of the Holling-Tanner model

"Model building is the art of selecting those aspects of a process that are relevant to the question being asked." - in: J.H. Holland, Hidden Order. 1995.

Consider the Holling-Tanner model (8.1). To begin the analysis we sketch the nullclines of (8.1), given by

$$(x' =) \quad 0 \;=\; r\left(1 - \frac{x}{K}\right)x - \frac{wx}{D+x}y, \qquad (8.4)$$
$$\text{and}$$
$$(y' =) \quad 0 \;=\; s\left(1 - J\frac{y}{x}\right)y.$$

The $x' = 0$ nullclines are the y-axis, $x = 0$, and a parabola opening down

$$y = \frac{r}{w}(D+x)\left(1 - \frac{x}{K}\right).$$

This parabola has x-axis intercepts (obviously) at $x = K$ and $x = -D$. The $y' = 0$ nullclines are the x-axis, $y = 0$, and a line with positive slope

$$y = \frac{s}{J}x.$$

The analysis that follows suggests that there are two natural cases to consider. The two nullclines have a unique fixed point (x^*, y^*) which can lie either before or after the local maximum of the $x' = 0$ null cline.

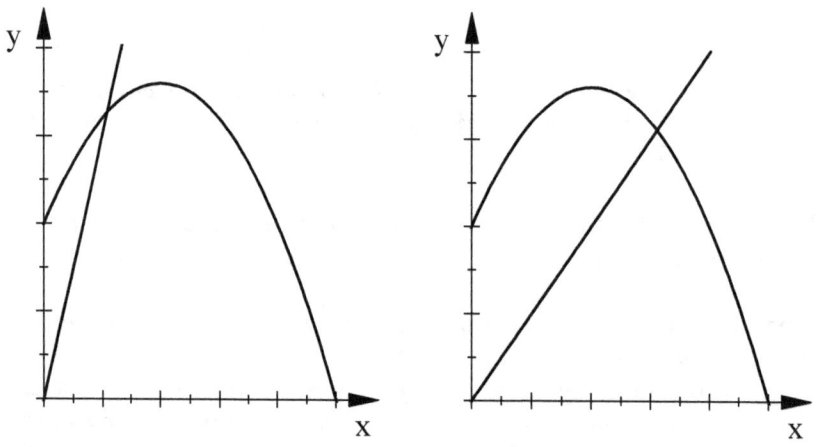

Null-clines: Cases 1, $x^* > (K - D)/2$, & 2, $x^* < (K - D)/2$

The pattern of trajectory directions, easily found by checking the signs of x', y', are the same in both cases. They are given next in Figure 8.3.

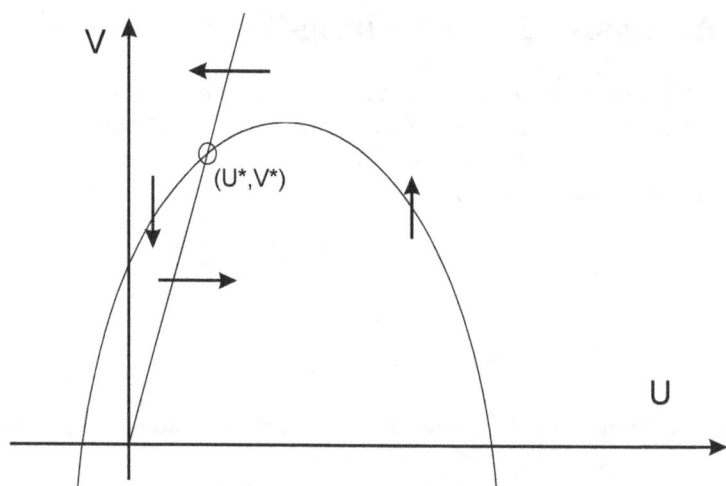

Figure 8.3: Trajectory directions

In both Cases 1 and 2 there is a positively invariant region, constructed in the next section, including the non-trivial equilibrium point. Further, in both cases the above trajectory directions imply that the equilibrium must be either a center or spiral. Thus, by the Poincaré-Bendixon Theory, *existence of a limit cycle depends on the stability of the included critical point.* A lengthy but straightforward bit of algebra reveals that the equilibrium point (x^*, y^*) is a stable spiral if

$$x^* > (K - D)/2,$$

that is, in Case 1.

We shall show however that in Case 2, (x^*, y^*) is unstable and a stable limit cycle thus exists by the Poincaré-Bendixon Theorem.

Theorem 54 *The Holling-Tanner model has a stable limit cycle if*

$$s < rx^* \frac{K - D - 2x^*}{K(x^* + D)}.$$

This theorem was first proven in the 1968 work of Tanner. In this section, we will also give a proof of the above result using Poincaré-Bendixon Theory. It is interesting that although the model has six(!) parameters the above theorem shows that its qualitative behavior is only governed by one combination of those six[2].

8.3.1 Rescaling the model

"Order and simplification are the first steps toward the mastery of a subject."
-Thomas Mann

Often models are so complex that it is difficult to perform analysis without repeated algebraic errors. In such cases, it is good practice to simplify by rescaling, called non-dimensionalization for models from mechanics, before performing analysis. This a fundamental step in applied mathematics. We illustrate it first for a simpler model before performing it on the Holling-Tanner model.

Example 55 *Rescaling the logistic model. The logistic model is*

$$N' = r(1 - N/K)N.$$

Rescale the variables by

$$
\begin{aligned}
u(t) &= \frac{N(t)}{K} \text{ (the fraction of carrying capacity)} \\
t' &= rt \text{ (a rescaled time).}
\end{aligned}
$$

Then the chain rule implies

$$
\begin{aligned}
\frac{du}{dt'} &= \frac{du}{dt} \times \frac{dt}{dt'} = \frac{d}{dt}\left(\frac{N(t)}{K}\right) \times \frac{1}{r} \\
&= \frac{1}{r}\frac{1}{K}N'(t) = \frac{1}{r}\frac{1}{K}r(1 - N(t)/K)N(t) \\
&= \left(1 - \frac{N(t)}{K}\right)\frac{N(t)}{K}.
\end{aligned}
$$

---

[2] There is a philosophical question of "What is mathematics?" One answer is that mathematics is a process for putting together numbers and pictures. Another, equally valid answer, is that mathematics, as happens here when it reveals that only 1 parameter is active, is a process for finding universal simplicity hiding in complexity.

Thus, $u(t)$ satisfies

$$u' = u(1 - u)$$

which is an equation with zero free parameters.

Since the Holling-Tanner model has exactly one non-trivial fixed point in the first quadrant (see the nullclines sketched previously) it is handy to rescale the equations by x^*. This also has the advantage of reducing the amount of book-keeping that would be necessary if we were to work with the original variables and parameters (compare (8.2,8.3) with the formula for (u^*, v^*) below!). Define the new, rescaled, variables and parameters as follows.

$$D^* = \frac{D}{x^*}, \quad K^* = \frac{K}{x^*}, \quad u(t) = \frac{x(t)}{x^*}. \quad v(t) = \frac{y(t)}{x^*}.$$

The rescaled equations are now given as follows.

$$\begin{cases} u' &= r\left(1 - \frac{u}{K^*}\right)u - \frac{wuv}{D^* + u} \\ v' &= s\left(1 - J\frac{v}{u}\right)v, \end{cases} \tag{8.5}$$

The fixed point (u^*, v^*) for this rescaled system (8.5) is now given by

$$(u^*, v^*) = \left(\frac{x^*}{x^*}, \frac{y^*}{x^*}\right) = \left(1, \frac{1}{J}\right).$$

8.3.2   Existence of a limit cycle

The rescaled model is

$$\begin{cases} u' &= r\left(1 - \frac{u}{K^*}\right)u - \frac{wuv}{D^* + u}, \\ v' &= s\left(1 - J\frac{v}{u}\right)v. \end{cases} \tag{8.6}$$

It has an equilibrium at

$$(u^*, v^*) = \left(1, \frac{1}{J}\right).$$

When the rescaled model (8.6) is linearized about the fixed point (u^*, v^*) the linearized system

$$\begin{cases} u' &= au + bv, \\ v' &= cu + dv \end{cases}$$

has the 2x2 linearization matrix A

$$A = \begin{bmatrix} a & b \\ c & d \end{bmatrix} = \begin{bmatrix} r\left(-\frac{1}{K^*} + \frac{w}{rJ(1+D^*)^2}\right) & -\frac{w}{1+D^*} \\ \frac{s}{J} & -s \end{bmatrix}. \tag{8.7}$$

The trace of a matrix, the sum of its diagonal entries, is equal to the sum of the eigenvalues of the matrix and the determinant is equal to their product. For a 2×2 matrix the trace and determinant combine to give all the information about the two eigenvalues.

Lemma 56 (Eigenvalues of 2×2 matrix) *Let*

$$A = \begin{bmatrix} a & b \\ c & d \end{bmatrix}$$

then the characteristic equation of A is

$$\lambda^2 - trace(A)\lambda + \det(A) = 0.$$

The eigenvalues of A are

$$\lambda_{1,2} = \frac{1}{2}trace(A) \pm \frac{1}{2}\sqrt{trace(A)^2 - 4\det(A)}.$$

The determinant of the linearization A is positive since a direct calculation gives

$$\det(A) = rs\left[\frac{1}{K^*} + \frac{wD^*}{rJ(1+D^*)^2}\right] > 0.$$

Since $\det(A) = \lambda_1\lambda_2$, the critical point (u^*, v^*) cannot be a saddle point.

Proof of Theorem 7.6. To investigate the eigenvalues of the above linearization matrix A, let us examine the trace and determinant of the linearization matrix A. The determinant of the linearization A is positive: a direct calculation gives

$$\det(A) = rs\left[\frac{1}{K^*} + \frac{wD^*}{rJ(1+D^*)^2}\right] > 0.$$

Since $\det(A) = \lambda_1\lambda_2$, as noted in the last section, the critical point (u^*, v^*) cannot be a saddle point. The trace of A is

$$trace(A) = r\left(\frac{w}{(rJ)(1+D^*)^2} - \frac{1}{K^*}\right) - s, \qquad (8.8)$$

which can be either positive or negative depending on the parameters' exact values. For example, if $trace(A) > 0$, $(u^*, v^*) = (1, \frac{1}{J})$ is an *unstable spiral* point. This occurs if and only if:

$$r\left(\frac{w}{(rJ)(1+D^*)^2} - \frac{1}{K^*}\right) > s \qquad (8.9)$$

which occurs if and only if (u^*, v^*) lies to the left of the maximum of the $u' = 0$ nullcline (we leave it as an exercise in algebra to verify this). When (8.9) fails, (u^*, v^*) lies to the right of the $u' = 0$ nullcline and, since in that case $trace(A) < 0$, (u^*, v^*) is a *stable spiral*.

Construction of a positively invariant region. We will focus on the case when (8.9) holds. In this case existence of a limit cycle is proven using the following strategy. We shall apply Poincaré-Bendixon Theory by finding an invariant region S which contains only the equilibrium point (u^*, v^*). This construction works for both cases 1 and 2 but when (8.9) does not hold the region contains a stable equilibrium. Since this equilibrium point is not stable when (8.9) holds it therefore will follow that the invariant region must contain a stable limit cycle.

First, consider the $u' = 0$ and $v' = 0$ nullclines, plotted in the next figure. Consider a trajectory which begins at the point $(K^*, 0)$. Looking at the nullclines and the trajectory directions, it is geometrically clear that this trajectory must rise and move to the left initially, then go down and to the left and then back up and to the right. In this manner it circles around the critical point (u^*, v^*) and then intersects the $u' = 0$ nullcline again at some point, let us call it p^*. This trajectory is illustrated in Figure 8.4.

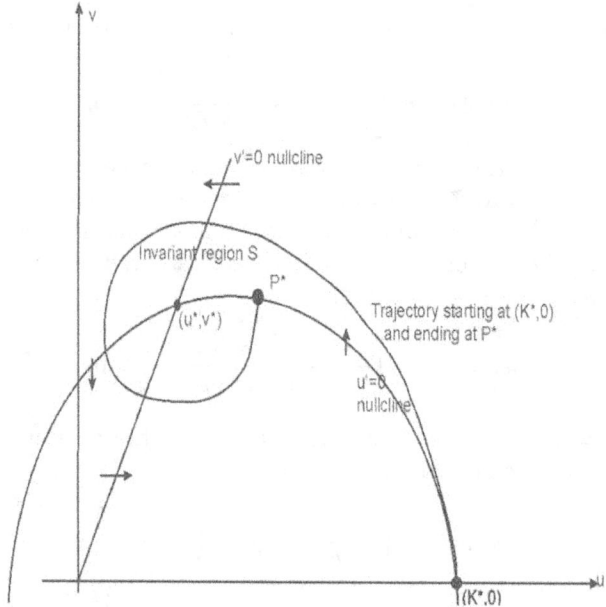

Figure 8.4: Illustration of an invariant region

Consider the region S bounded by a piece of this trajectory, as depicted in Figure 8.4, and a piece of the nullcline $u' = 0$. Trajectories are tangent to one part of the boundary of S (the part being a trajectory) and enter along the other part of the boundary of S. Thus S is an invariant region for the Holling-Tanner system. When (8.9) holds the equilibrium point is not stable so, by Poincaré-Bendixon Theory, there must be a limit cycle in the region S. Thus, we have proven Tanner's existence theorem, Theorem 54.

To summarize, there were two features of the proof: existence of a positively invariant region around the equilibrium point (in both cases) and stability (case 1) or instability (case 2) of the enclosed equilibrium point.

8.4 Bifurcations in Holling-Tanner

"What was scattered gathers. What was gathered blows away." - *Heraclitus*

From the analysis of the previous section, we know that the Holling-Tanner model has a positive invariant region with only one equilibria in the region. The pattern of trajectory directions around that equilibrium eliminates the possibility that the equilibrium is a node or saddle. Thus the equilibrium can only be a center or a stable spiral or an unstable spiral. Starting with this geometric observation reduces the possibilities that the analysis must consider.

In this section we apply the Hopf bifurcation theorem to (8.1) to study the limit cycles described in the previous section by Poincaré-Bendixon Theory. Recall from the previous section that the determinant of the linearization matrix A is always strictly positive while the trace of the matrix A can be either positive or negative:

$$\det(A) > 0$$

but

$$trace(A) = r\left(\frac{w}{rJ(1+D^*)^2} - \frac{1}{K^*}\right) - s \tag{8.10}$$

can have two signs.

Since $det(A)$ is positive, (x^*, y^*) can never be a saddle point. This is also clear from the pattern of trajectories.

To search for the precise point at which the limit cycle is born, we seek the point where the eigenvalues of A cross the imaginary axis. At this precise point $trace(A)$ will vanish. Thus, setting $trace(A) = 0$ gives the parameter value

$$r\left(\frac{w}{rJ(1+D^*)^2} - \frac{1}{K^*}\right) - s = 0. \tag{8.11}$$

It is interesting that (8.11) suggests that the one combination of the six parameters give by the right hand side of (8.11) is the one natural parameter of the Holling-Tanner system (8.1).

To simplify our description somewhat let us regard all of the parameters except one as fixed and view one parameter in (8.1) as a distinguished parameter. (Of course, the actual distinguished parameter is the above precise combination.) For example, let us regard $r, J, K, w,$ and D as fixed and study (8.1) as the reproduction rate of the predators, s, varies. Recall that s is a reproduction rate and thus positive. To have the critical value of s

$$s_{crit} := r\left(\frac{w}{rJ(1+D^*)^2} - \frac{1}{K^*}\right) \tag{8.12}$$

positive the carrying capacity K must be sufficiently large. We make the following assumption on the model parameters.

Assumption A4. K is sufficiently large that

$$s_{crit} = r\left(\frac{w}{rJ(1+D^*)^2} - \frac{1}{K^*}\right) > 0.$$

Supposing **Assumption A4** holds, the fact that the trace of A changes sign at $s = s_{crit}$ implies that

- For $s > s_{crit}$, (x^*, y^*) is an asymptotically stable critical point.

- For $s < s_{crit}$, (x^*, y^*) is an unstable spiral.

Thus, a stable limit cycle will emerge as s decreases through s_{crit} provided (x^*, y^*) is asymptotically stable when $s = s_{crit}$. We will leave it as an exercise to verify that when $s = s_{crit}$ the equilibrium point is asymptotically stable. Thus the Holling-Tanner model undergoes a Hopf bifurcation as s passes through this critical value s_{crit}.

Remark 57 *Since $K \uparrow$ implies $s_{crit} \downarrow$, increasing K can destabilize the equilibrium causing population oscillations. This is sometimes called the paradox of enrichment. In extreme cases it can lead to overshoot and collapse of populations.*

8.5   Testing the model

"To test a recently developed predator-prey model against reality, I chose the well-known Canadian hare-lynx system. ... The correlation between the model and the empirical data gives some idea about the general worth of the model..... But the regression fit was equally poor. In fact it was worse than poor; it was impossibly bad." - M.E. Gilpin, 1973.

There have been numerous attempts to take populations of predators and their primary prey, find the model parameters and compare model predictions with actual population levels. Generally, predator prey models of all types have proven to be successful in explaining qualitative behavior of populations but disappointing in quantitative predictions compared for real populations. As an example of qualitative success, Tanner in 1968 studied data from a number of predator-prey pairs. He found that the qualitative predictions of the Holling–Tanner model closely correlated with whether the populations cycled or were (very nearly) at equilibrium. This is a significant success of the Holling-Tanner model!

However, the Holling-Tanner models fails at accurate prediction in a similar way the Lotka-Volterra model does. In 1963, Rosenzweig and McArthur analyzed predator-prey data by plotting it in the phase plane for the first time. In 1973, Gilpin plotted the actual Lynx - Snowshoe hare data in the phase plane. We have replotted Gilpin's data in the same axes below in Figure 8.5. The data indicates a clear clockwise cycle. Both the Holling-Tanner model and the Lotka-Volterra model predict a counterclockwise cycle. Thus, the data fits the model only if hares are the predator and lynx the prey! Obviously, the Holling-Tanner model is an oversimplification that is useful for understanding some features of interacting populations but leaves other features unexplained.

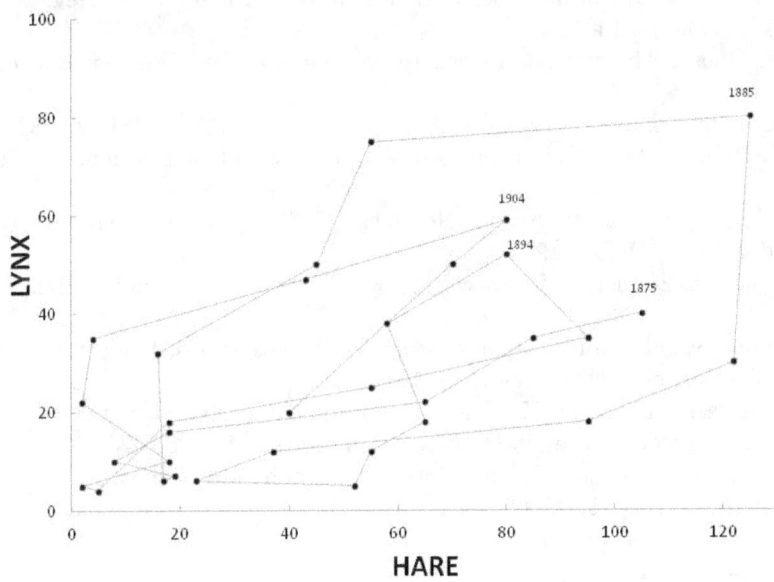

Figure 8.5: The Lynx-Hare data plotted. Note general clockwise cycle.

8.6 References for Chapter 8

"The above formulae constitute the basic units in a "build-a-model" toy. They may be assembled in various combinations to give a one-predator-one-prey model considerably more sensible than the simple Lotka-Volterra one." - R. May, p. 84 in: Stability and Complexity in Model Ecosystems

We have seen that the deficiencies in the Lotka-Volterra model have spurred a rich interaction of mathematicians and ecologists. This has yielded a greater understanding of how populations interact and a rich development of mathematical realizations of that understanding. The historical tradition in science, going back to John Locke, had been to isolate components in a system and view system behavior as a combination of component behavior. One contribution of Lotka-Volterra and subsequent interacting population models has been to show that system behavior can also arise from interactions that cannot be studied in isolation. The Holling-Tanner model shows a clear example of the mathematical development of such an extended model.

References

V.I. Arnold, Ordinary Differential Equations, Massachusetts Institute of Technology Press, Cambridge, Massachusetts, 1973.

J. Cronin, Some mathematics of biological oscillations, SIAM Review, volume 19 (1977), 100-138.

G.F. Gause, The struggle for existence, Williams and Wilkins, Baltimore, 1934.

M.E. Gilpin, Do hares eat lynx?, American Nat. v.107 (1973) 727-730.

C.S. Holling, The components of predation, Canadian Entomol. 91(1959) 293-320.

D.W. Jordan, and B. Smith, Nonlinear Ordinary Differential Equations, Clarendon Press, Oxford 1977.

A.J. Lotka, Elements of physical biology, Williams and Wilkins, Baltimore, 1925.

R. May, Stability and Complexity in Model Ecosystems, Princeton University Press, Princeton New Jersey, 1974.

M.L. Rosenzweig and R.K. McArthur, Graphical representation and stability conditions of predator-prey interactions, Am. Natur. 47(1963) 209-223.

J.T. Tanner, The stability and intrinsic growth rate of prey and predator populations, Ecology, 56 (1968), 855-867.

8.7   Exercises for Chapter 8

1. Choose values for r, s, w, K, D, and J satisfying $s < s_{crit}$. Plot the resulting phase portrait.

2. If logistic growth is replaced by a pure birth process in Assumption A1, does the resulting system admit a limit cycle? Give both analytic reasoning for your answer and one example of a plotted phase portrait.

3. In the construction of a positively invariant region for the Holling-Tanner model, what prevents the trajectory beginning at $(K^*, 0)$ from hitting the $u = 0$ axis?

4. Consider the model

$$
\begin{aligned}
x' &= x(1 - y - x^2), \\
y' &= y(-\frac{1}{2} + x - y^2).
\end{aligned}
$$

   Plot the nullclines, equilibria and trajectory directions. Find a positively invariant region. Analyze stability of the equilibria within it and draw conclusions.

5. Consider the rescaled Holling-Tanner model

$$
\begin{cases}
u' &= r\left(1 - \frac{u}{K^*}\right)u - \frac{wuv}{D^*+u} \\
v' &= s\left(1 - J\frac{v}{u}\right)v,
\end{cases}
$$

   with critical point $(1, 1/J)$. Determine its stability when $s = s_{critical}$, either by analysis or by careful numerical experiments.

6. Show that if additionally we rescale time $t' = st$ then one additional parameter is eliminated from the model.

7. Consider a predator prey interaction satisfying the 3 assumptions below.

A1: The prey undergo a pure birth process in the absence of the predator while the predator undergoes a pure death process in the absence of the prey.

A2. Interactions between the two increases the growth rate of the predator and decreases the growth rate of the prey in proportion to the number of interactions.

A3. Crowding and competition among predators decreases the growth rate of the predators in proportion to the number of interactions among predators.

Write down an ODE model of this process.

Chapter 9

Business Cycles

9.1 Introduction

"Business-cycle theorists concerned themselves with why the economy naturally generated fluctuations in employment and output, [while the rest of the profession] continued to operate on the assumption that full employment was the natural, equilibrium position for the economy." - Robert Aaron Gordon, Business Fluctuations (1952), p. 340

Many models of economies are static: an "optimal" economic equilibrium describing distribution of goods, services, labor and capital is sought. This chapter develops a mathematical model of cycles as an essential feature of economic activity. The model indicates that business cycles *can* occur due to intrinsic economic relationships in addition to normal, periodic fluctuations, such as seasonal variations in productivity. Classical economic theories postulate that rational decisions are made which force the economy to approach steadily that optimal state. On the other hand, economies fluctuate, and business cycles are not solely in response to seasonal changes. The current state of business cycle models is not to predict

Will it happen?

Their goal is rather to identify principles that can answer

Could it happen for a simple reason?

As such, business cycle theories cannot produce predictions but may identify key variables for further study.

9.2 Business cycle theories

"The researches of many commentators have already thrown much darkness on this subject, and it is probable that if they continue we shall soon know nothing at all about it." -Mark Twain

Many types of economic cycles have been described by different economists. These include long wave theories, investment/ business cycle theories, random fluctuations, cycles due to debt and exogenous cycles caused by cyclic externalities. We review a few of these.

Long Wave Theories.

There are (at least) four theories about long wave economic cycles. The most famous is the **Kondratieff capital-investment theory** due to the economist Kondratieff (1892 to about 1930). This argues that long term cycles occur due to investment in infrastructure, the growth that is caused by improved infrastructure followed by decline due to its depreciation and decrepitude. The **capital crisis theory** of Leon Trotsky argued that crises and declines are due to declines in profit rates. The **war theory** argues (with supporting correlations) that long wave economic cycles are due to recurring major wars. The **innovation theory** of long waves argues that long period cycles are caused by growth due to clusters of innovations followed by structural change, followed by saturation, followed by stagnation and then further technological innovation. These are also called **technological innovation cycles**. The description of **long wave innovation cycles** in e.g. Perez 2002 (for example Figure 3.1 p. 30) is similar to describing population growth as a series of S-curves (called logistic *escalation*). If the production undergoes logistic escalation and the production rate $P'(t)$ is plotted then a clear wave like pattern of *Expansion, Peak, Contraction, Trough* emerges.

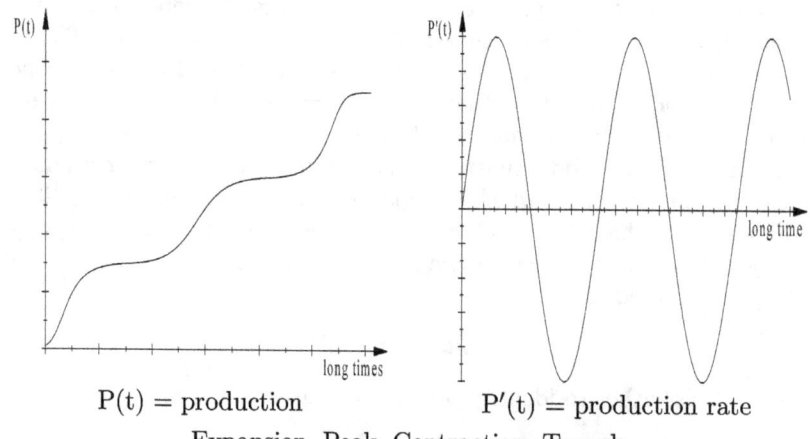

$P(t)$ = production $P'(t)$ = production rate

Expansion, Peak, Contraction, Trough

A standard long wave theory would put historical peaks' years and reasons for the peaks as follows.

1825	1873	1913	1956	2000
Steam Eng.	*Rail*	*Chemistry*	*PetroChems*	*Internet*
Cotton gin	*Steel*	*Elec_Eng*	*Autos*	*Computers*

These long waves are *cyclic but not periodic* as the cycle length, the time from peak to peak, varies.

Investment cycles.

These are medium term oscillations and are the type of cycles described by most business cycle models, like Goodwin's model consider here. The economist Juglar (1819 to 1905) observed 7 to 11 year long cycles in investment rates, prices, banking, inflation and employment. Early data systematically available detailing cyclic fluctuations was codified in a chart called *the Harvard Barometer*, developed by Harvard economist Warren Milton Persons[1] (1878–1937) . An example plotted from the data[2] appears below in Figure 9.1.

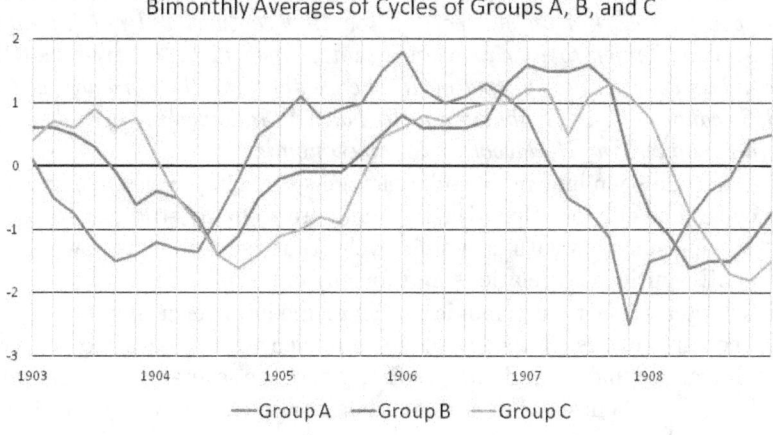

Cycles in Bi-Monthly Averages
January 1904 to July 19114
Figure 9.1:   GROUP A = INDUSTRIAL STOCKS AND RAILROAD BONDS
GROUP B = LABOR PRICES AND BANK RESERVES
GROUP C = RATE OF FOUR TO SIX MONTH PAPER

Cycles due to random fluctuations.

Short term cycles of between two and four years were reported by the economist Kitchin (1926). These cycles are typically described as short term fluctuations due to randomness in economic activity.

Cycles associated with Minsky's "Financial Instability Hypothesis".

Minsky described boom and bust cycles due to debt and optimism, both factors ignored in classical economics. These proceed as follows. Low debt and growth stir optimism. Over optimism leads to borrowing to invest for further

---

[1] Warren Milton Persons, 1878-1937, was an economist at Harvard until he left in 1928. He created the cycle indicator called the "Harvard barometer" in 1919. The barometer was influential but ended when Persons left Harvard.

[2] The original figure is done in pen and ink by an artist on page 112 in: W.M. Persons, Indices of General Business Conditions, Harvard Univ. Committee on Economic Research, 1919. This is a dry approximation.

growth. Eventually debt expands so that small shocks mean debt can not be serviced. This leads to collapse followed by low debt and the cycle repeats.

Cycles due to external factors: Sunspots, Animal Spirits, · · ·.

"Azariadis (1981) was the first published paper to show that sunspots may be responsible for business cycles although he uses the term self-fulfilling prophecies, originally coined by Robert K. Merton (1948). Woodford (1986, 1988) further demonstrated how sunspots could be relevant to understanding macroeconomic fluctuations. Howitt and McAfee (1992) use the term 'animal spirits' (popularized by Keynes in the General Theory) to refer to the same concept. It is perhaps unfortunate that these terms are now closely connected. Jevons, for example, who worked on sunspots in the 19th century, did not intend that 'sunspots' should refer to extrinsic uncertainty; instead he believed that there was a real link between the sunspot cycle, the weather and the agricultural sector of the US economy." -J. Benhabib and R. Farmer, Indeterminacy and sunspots in macroeconomics, in: Handbook of Macroeconomics

These include spending in periodic elections, growing cycles, sunspots and climate cycles like el Nino. Surprisingly, there are so many studies and pseudo-studies[3] linking sunspots with economic cycles that "sunspots" is now a generic term for both spurious correlations and for any externalities.

This chapter presents a dynamic model of medium term investment, business, or economic cycles. These cycles have proven to be as easy to observe as they are difficult to understand and predict. There are several problems that immediately present themselves in studying business cycles. First, generally there are no true cycles; rather one observes cycles about a mean which generally is growing. Second, the business environment is complex and subject to seasonal (periodic) input. Thus it can be hard to separate periodic responses to periodic inputs (seasonal or demographic cycles) or forces from periodic fluctuations due to the inherent internal dynamics of economies.

Before developing a model of business cycles we first review some of these difficulties.

9.2.1   Some basic difficulties

No two business cycles are quite the same; yet they all have much in common. They are not identical twins, but they are recognizable as belonging to the same family." - Paul Samuelson, Economics (2nd ed., 1951), Ch. 18 : The Business Cycle

"Interesting phenomena occur when two or more rhythmic patterns are combined, and these phenomena illustrate very aptly the enrichment of information that occurs when one description is combined with another." - Gregory Bateson

It is undeniable that good and bad economic periods alternate. Thus it seems easy to simply accumulate data on various industries, identify alternating patterns and then look for regularity in the patterns that can be captured in a mathematical model. Unfortunately there are severe problems with this

---

[3] Try an internet search for "sunspots and business cycles"!

program. First, (see the Harvard Barometer figure) the good and bad times can be industry dependent, fluctuate wildly due to noise, seasonal influences, sunspots and so on. Further, mathematical cycles as exactly periodic oscillations just do not occur. Markets move up and down with time scales that vary, unlike the tides! Business cycles can only be considered qualitatively similar to another, not quantitatively similar, as pointed out by the Economist W.C. Mitchell (1874-1948). While precipitating events are often discussed in the finance industry, the actual causes are certainly deeper and hidden from view.

Identifying cycles in data can be hard due to quasi-periodicity. Even supposing exact T−periodicity, both data and measurements are imprecise so a periodic function observed (with measurement errors) will never satisfy $|f(t+T)-f(t)| = 0$. Thus it is reasonable to quantify the errors in the data, ε, and ask instead if the function is periodic to within data errors:

$$|f(t+T) - f(t)| < \varepsilon.$$

If we accept the description of economic cycles of short term 40 month ($3.33\overline{3}$ years) cycles (Kitchen), medium term 9 year cycles (Juglar) and long term 60 year cycles (Kondratieff) then economic activity is at minimum a superposition (sum) of 3 periodic functions with different periods ω_j (e.g., $\omega_j = \frac{10}{3}, 9, 60$) such as

$$f(t) = A\cos\frac{2\pi t}{\omega_1} + B\cos\frac{2\pi t}{\omega_2} + C\cos\frac{2\pi t}{\omega_3}.$$

An example that is as simple as possible, with only 2 frequencies, is plotted in Figures **??**, **??** below.

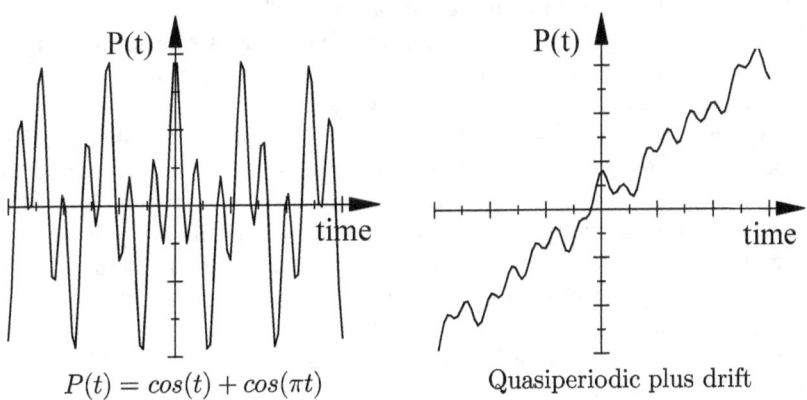

$P(t) = cos(t) + cos(\pi t)$ Quasiperiodic plus drift

A function that is the sum of periodic functions with incommensurable periods is called quasi-periodic.

Definition 58 *Two periods are incommensurable if their ratio $\frac{\omega_1}{\omega_2}$ is an irrational number.*

A quasi-periodic function is the sum of periodic functions with (a finite number of) two or more incommensurable periods.

T = T(ε) is an ε-almost period if the function is periodic to within a tolerance ε, specifically

$$|f(t+T) - f(t)| < \varepsilon, \text{ for all } t.$$

It is known that a quasi-periodic function is also almost periodic and thus satisfies the following property of Bohr.

Theorem 59 (Bohr) *A quasi-periodic function has the property that, for any ε > 0, and any interval length l > 0 the interval [ω, ω + l] contains an ε-almost period.*

Thus, since economic data will have measurement errors and finite precision, if it is quasi-periodic, the analysis of the data will appear to show that it is periodic of any period testable within measurement error. For this and many other reasons, the search for a fixed period in business activity is hopeless.

9.3   Goodwin's model

"Economists also use models to learn about the world, but instead of being made of plastic, they are most often composed of diagrams and equations. Like a biology teacher's plastic model, economic models omit many details to allow us to see what is truly important. Just as the biology teacher's model does not include all the body's muscles and capillaries, an economist's model does not include every feature of the economy." - N. Gregory Mankiw, Principle of Economics (6th ed., 2012), Ch. 2. Thinking Like an Economist

Observed business cycles pose a puzzle. The obvious solution is simply to assert that all cycles are caused by cyclic inputs (sunspot / animal spirits). It is impossible to set up an experiment with two economies in bottles isolated from time varying influences. Goodwin studied *"Could it happen?"* using a different and compelling approach. He wrote down a simplified mathematical model (a cartoon) of an economy that captures a simplified version of the classical economics. He then interrogated the model asking if periodic or cyclic solutions were intrinsic in solutions rather than caused by a periodic model input. Goodwin showed a simple, two compartment model of an economy which exhibited oscillations similar to predator prey oscillations due to the inherent interactions of economic variables. The oscillations predicted by Goodwin's 1967 model can be described in words as follows

If wages are low, a higher fraction of production is invested thereby increasing output.

Higher output requires higher employment.

This puts pressure upon wages; rising wages increase the labor bill share which decreases investment and thereby decreases overall productivity.

Employment thus decreases.

This results in lower wages and the cycle begins anew.

9.3.1 The Variables

"Business cycles are a type of fluctuation found in the aggregate activity of nations that organize their work mainly in business enterprises: a cycle consists of expansions occurring at about the same time in many economic activities, followed by similarly general recessions, contractions, and revivals which merge into the expansion phase of the next cycle; this sequence of changes is recurrent but not periodic; in duration business cycles vary from more than one year to ten or twelve years; they are not divisible into shorter cycles of similar character with amplitudes approximating their own." - Arthur F. Burns and George W. Mitchell (1946, 3); as cited in: Robert J. Gordon, ed. The American Business Cycle: Continuity and Change, 1986. p. 1

The first difficulty is that economic activity is really a superposition of cyclic behavior plus drift. Somehow the growth / decay / drift must be filtered out in the variables selected to study cycles. Typically this is done by either normalizing the variables by their mean or by subtracting the mean. The model of Goodwin involves two normalized variables:

$$e := \text{employment rate} = \frac{\text{number employed}}{\text{total labor supply}},$$
$$\text{and}$$
$$\Theta := \text{labor bill share} = \frac{\text{total workers income}}{\text{total production}}.$$

e and Θ are thus normalized to lie between 0 and 1:

$$0 \leq e \leq 1$$
$$\text{and}$$
$$0 \leq \Theta \leq 1.$$

Goodwin's model considers the relationship between capital, labor and income and is a mathematical description of the principles of classical economics. Specifically, it assumes that *all markets respond instantly to prevailing conditions*: there is no inertia or stickiness in, for example, wages going up or down. Another idealized assumption is that *savings equals investment*. Although we will end with a system of two equations for the variables e and Θ, we will need a number of other variables along the way. Define the following parameters.

$w :=$ average wage,

$N(t) :=$ number of employed workers,

$N^* :=$ total number of workers,

$Y :=$ total national output or production,

$e :=$ employment rate $= \frac{N(t)}{N^*} = \frac{\text{number of employed workers}}{\text{total number of workers}}$,

$P :=$ labor productivity $= \frac{Y}{N(t)} = \frac{\text{total national output}}{\text{number of employed workers}}$
and
$K :=$ total available capital.

The labor bill share can be expressed in terms of these variables as

$$\Theta = \frac{\text{wage} \times \text{total employed}}{\text{total output}} = \frac{wN}{y}.$$

9.3.2   The Assumptions of Goodwin's Model

"Knowledge of terms leads to knowledge of things." - Plato.

The first assumption that we need to develop the model is that the rate of increase of wages is tied to the employment rate. As full employment nears the wage required to add one more worker increases and at full employment it is impossible to hire another worker, expressed as the wage required is infinite: $f(e) \to \infty$ as $e \to 1$.

Assumption A1 (employment rate determines wages): *The relative rate of increase of wages is a function of the employment rate:*

$$\frac{w'}{w} = f(e), \tag{9.1}$$

where

$$f(0) < 0, \ f'(e) > 0 \text{ and } f(1) = \infty.$$

The function $f(e)$ is called the *production function*. Its general shape is typically represented as in Figure 9.2. The above properties of the production function imply that its linear approximation takes the form.

$$f(e) = -a + be, \tag{9.2}$$

for some constants a and b. Goodwin's model assumes that $f(e)$ is given by its linear approximation (9.2) above, Figure 9.3.

Assumption A2 (linear approximation to $f(e)$): *The production function is given by $f(e) = -a + be$.*

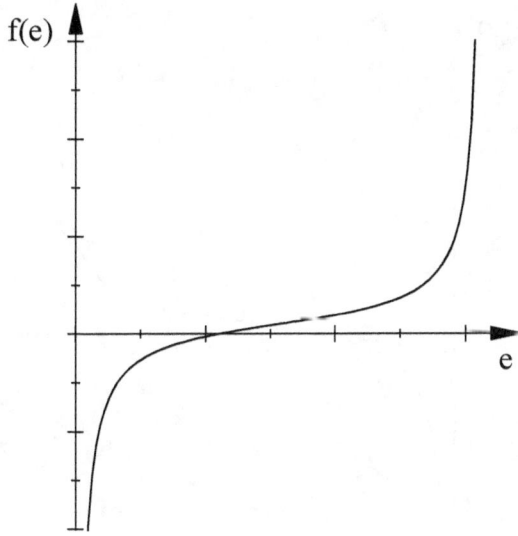

Figure 9.2:   Production Function:
              wage growth $= f(e)$

Assumption A2 can be modified to a production function with asymptotes at $e = 0$ and $e = 1$ as depicted above.

Recall that the labor productivity, P, is the output per worker:

$$P := \text{labor productivity} = \frac{Y}{N(t)} = \frac{\text{total national output}}{\text{number of employed workers}}.$$

Assumption A3 (Constant growth rates): *The following two growth assumptions hold*

The number of workers is growing at a constant rate:

$$\frac{(N^*)'}{N^*} = n, \qquad n > 0$$

The labor productivity is also growing at a constant rate[4]:

$$\frac{P'}{P} = d, \qquad d > 0.$$

Assumption A3 can obviously be modified to logistic growth.

---

[4] This assumption can only be valid over moderate time intervals. Over long time intervals, diffusion of technological innovations will cause intermittent large jumps in productivity. Wars and epidemics will simiarly cause intermittent large drops in productivity. Wars, epidemics and technological innovations are external factors. Since the question business cycle models try to answer is whether cycles can occur due to internal market dynamics alone, external factors are ignored in the models.

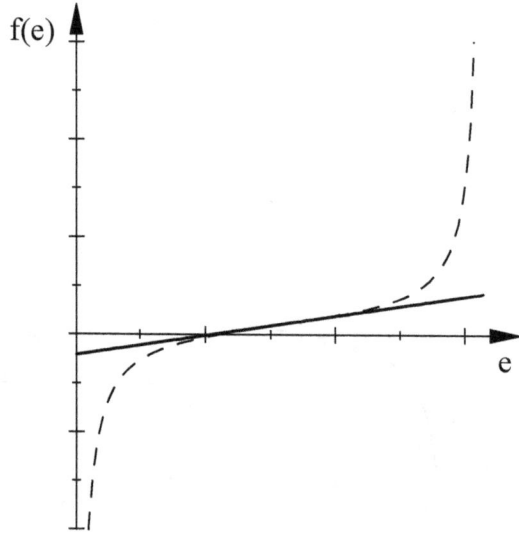

Figure 9.3: f(e) & **f(e) = linear approximation**

Assumption A4 (Output $\propto$ Capital): *The capital to output ratio is constant:*

$$\frac{K}{Y} = r, \qquad r > 0.$$

The following assumption will be used to relate $\frac{K'}{K}$ to e and Θ, thereby closing the system.

Assumption A5 (Savings = Investment): *All production not spent in workers' wages is invested, not consumed, and thus increases capital:*

$$K' = Y - wN. \tag{9.3}$$

This can be modified to some fraction of workers' wages invested and some fraction of capital consumed.

9.3.3   Derivation of Goodwin's Model

Economic activity is naturally phrased in terms of relative growth rates. We thus begin with an observation on relative rates of change.

Lemma 60 (Quotient rule lemma) *If*

$$u(t) = \frac{N(t)}{D(t)}$$

then

$$\frac{u'(t)}{u(t)} = \frac{N'(t)}{N(t)} - \frac{D'(t)}{D(t)}.$$

Proof. This is a calculation with the quotient rule. ∎

Since $e = \frac{N}{N^*}$ and $\Theta = \frac{wN}{Y}$ the quotient rule lemma implies

$$\frac{e'}{e} = \frac{N'}{N} - \frac{(N^*)'}{N^*}$$

and

$$\frac{\Theta'}{\Theta} = \frac{(wN)'}{wN} - \frac{Y'}{Y}.$$

Expanding this gives

$$e' = \left[\frac{N'}{N} - \frac{(N^*)'}{N^*}\right] e = [I - II] e$$

and

$$\Theta' = \left[\frac{(wN)'}{wN} - \frac{Y'}{Y}\right] \Theta = [III - IV]\Theta$$

where

$$I := \frac{N'}{N} \quad , \quad II := \frac{(N^*)'}{N^*},$$

$$III := \frac{(wN)'}{wN} \quad , \quad IV := \frac{Y'}{Y}.$$

To close the system we must therefore express the four terms on the above right hand sides, I, II, III, IV, in terms of e and Θ. We consider these four terms one by one in the order II,IV,I, III. To close the model each must be expressed in terms of the model's variables.

Term II $= \frac{(N^*)'}{N^*}$: By the growth Assumption A3 we have

$$II = \frac{(N^*)'}{N^*} = n, \text{ a constant.}$$

Term IV $= \frac{Y'}{Y}$: The assumptions on y are that

$$K = rY, \text{ and } K' = Y - wN.$$

Thus, since $K' = ry'$ we have

$$rY' = K' = Y - wN$$

so that

$$\frac{Y'}{Y} = \frac{1}{r}\left(1 - \frac{wN}{Y}\right) \text{ and } \Theta = \frac{wN}{Y}$$

so there follows

$$IV = \frac{Y'}{Y} = \frac{1}{r}(1 - \Theta).$$

Term I $= \frac{N'}{N}$: Recall that the worker productivity is $P = \frac{Y}{N}$ so that (again by the quotient rule lemma and the growth assumption A3)

$$d \;=\; \frac{P'}{P} = \frac{Y'}{Y} - \frac{N'}{N}$$

which yields

$$d \;=\; IV - \frac{N'}{N}.$$

Thus, from the analysis of term IV

$$d \;=\; \frac{1}{r}(1 - \Theta) - \frac{N'}{N}$$

rearranging

$$I \quad : \quad = \frac{N'}{N} = \frac{1}{r}(1 - \Theta) - d.$$

Term III $= \frac{(wN)'}{wN}$: Expanding term III gives

$$\frac{(wN)'}{wN} \;=\; \frac{wN' + w'N}{wN} = \frac{w'}{w} + \frac{N'}{N}$$

$$\;=\; \frac{w'}{w} + \text{term } I = \frac{w'}{w} + \frac{1}{r}(1 - \Theta) - d.$$

The assumption linking wage growth to employment rate is $w'/w = f(e)$. Thus

$$III = -a + be + \frac{1}{r}(1 - \Theta) - d.$$

The Model. To summarize we have the following

$$I = \tfrac{1}{r}(1 - \Theta) - d, \qquad\qquad II = n$$

$$III = -a + be + \tfrac{1}{r}(1 - \Theta) - d, \qquad IV = \tfrac{1}{r}(1 - \Theta).$$

These are inserted into the model

$$e' \;=\; [I - II]\,e$$
$$and$$
$$\Theta' \;=\; [III - IV]\Theta.$$

This yields the **Goodwin model of business cycles**

$$e' = \left[\frac{1}{r}(1 - \Theta) - d - n\right] e$$

$$\& \qquad\qquad\qquad \text{(Goodwin Model)}$$

$$\Theta' = [-a - d + be]\Theta.$$

where r, d, n, a and b are all positive constants. This system bears a remarkable resemblance to the predator prey system from Chapter 4.

9.4   Analysis of the Goodwin Model

"When things go badly, people become cautious. Then their caution causes things to go well, and when things go well, they become incautious. I think that's a forever cycle."- Howard Marks

Compare the business cycle model with the predator prey model (for rabbits $R(t)$ and Foxes $F(t)$)

$$
\begin{array}{ccc}
e' = \left[\left(\frac{1}{r} - (d+n)\right) - \frac{1}{r}\Theta\right] e & , & R' = [a - bF]R \\
and & \Leftrightarrow & and \qquad \text{(Cycle vs LV)} \\
\Theta' = [-(a+d) + be]\Theta & , & F' = [-n + mR]F
\end{array}
$$

It is clear that this is exactly the same model (with redefined parameters) when $\frac{1}{r} - (d+n) > 0$. Thus the analysis from the Chapter on the Lotka-Volterra model implies the following.

Theorem 61 *If*

$$
\frac{1}{r} > (d+n)
$$

then Goodwin's model predicts periodic solutions that oscillate about the equilibrium (e^*, Θ^*)

$$
e^* = \frac{a+d}{b}, \quad \Theta^* = 1 - r(d+n).
$$

The time averages of $e(t)$ and $\Theta(t)$ over one period are exactly the equilibrium values.

Examples of business cycles in this case are displayed next, Figure **??**.

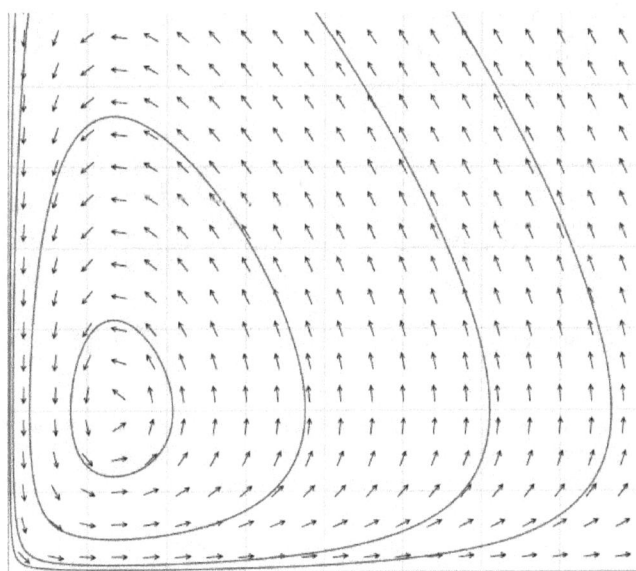

Lotka-Volterra type business cycles: $e - \Theta$ phase plane

9.4.1   When $\frac{1}{r} < (d + n)$

We now turn to the case where $\frac{1}{r} < (d+n)$. Redefining parameters, the system has the following form

$$e' = -\alpha e - \beta e\Theta,$$
$$\text{and}$$
$$\Theta' = -\gamma\Theta + \delta e\Theta,$$
$$\text{where}$$
$$\alpha = (d+n) - \frac{1}{r} \quad , \quad \beta = \frac{1}{r}$$
$$\gamma = a + d \quad , \quad \delta = b.$$

In the chart of human interaction models from Chapter 4, this corresponds to D - D and C - K which states that:

Both x and y undergo a death process in the absence of the other.

x cooperates with y and y competes with x.

From this interpretation, we all hope that $\frac{1}{r} < (d+n)$ is not the expected case. The phase plane has nullclines

$$e' = 0 \quad \Leftrightarrow \quad e = 0 \quad or \quad \Theta = -\frac{\alpha}{\beta}$$
$$\text{and}$$
$$\Theta' = 0 \quad \Leftrightarrow \quad \Theta = 0 \quad or \quad e = +\frac{\gamma}{\delta}.$$

These nullclines and trajectory directions are given in the next two figures 9.4 and 9.4.

Typical trajectories forecast growth followed by an inevitable crash. The phase plane has nullclines

$$e' = 0 \quad \Leftrightarrow \quad e = 0 \quad or \quad \Theta = -\frac{\alpha}{\beta}$$
$$\text{and}$$
$$\Theta' = 0 \quad \Leftrightarrow \quad \Theta = 0 \quad or \quad e = +\frac{\gamma}{\delta}.$$

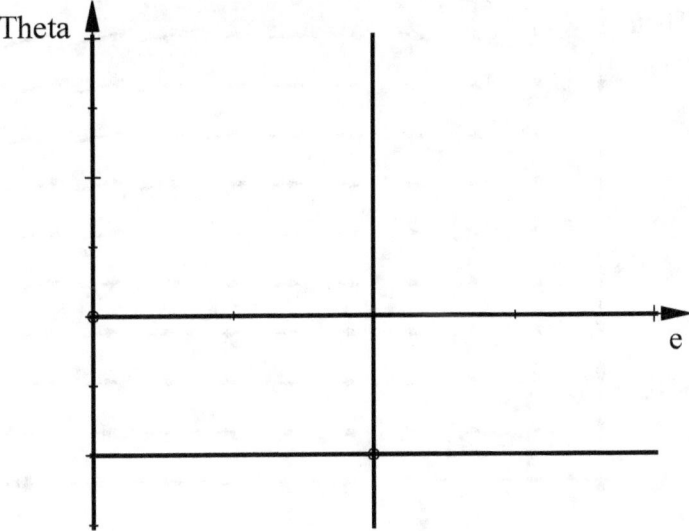

Figure 9.4: Nullclines when $\Theta^* < 0$

trajectory directions and a few representative trajectories are depicted below.

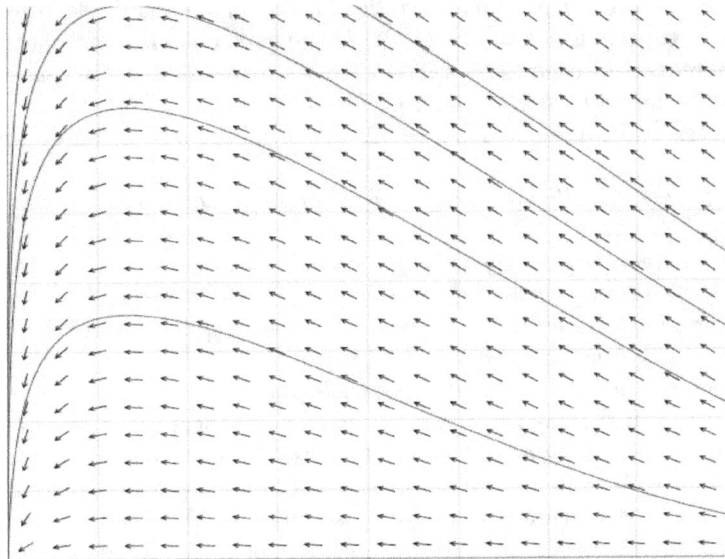

Business cycle collapse: $e - \Theta$ phase plane

Goodwin's model in the case when $\frac{1}{r} < (d + n)$ predicts a total collapse of business activity.

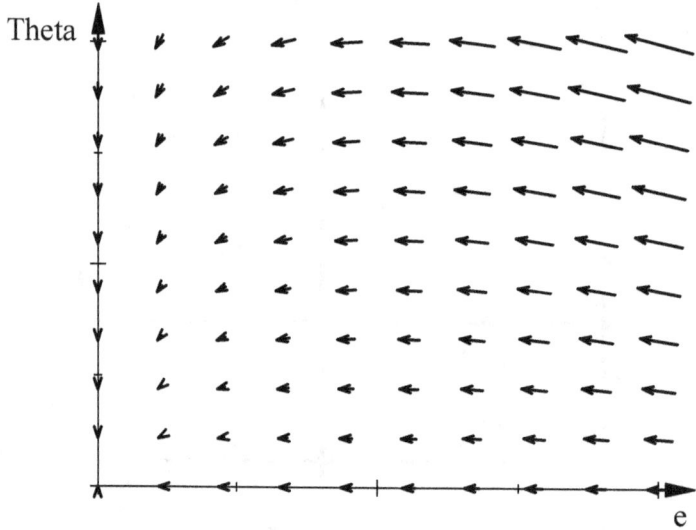

Figure 9.5: Trajectory directions when $\Theta^* < 0$

9.5　Conclusions from a simple model

My decision to leave applied mathematics for economics was in part tied to the widely-held popular belief in the 1960s that macroeconomics had made fundamental inroads into controlling business cycles and stopping dysfunctional unemployment and inflation. - Robert C. Merton

The oscillations predicted by Goodwin's 1967 model can be described in words as:

If wages are initially low, a higher fraction of production is invested thereby increasing output.

Higher output requires higher employment.

This puts pressure upon wages; rising wages increase the labor bill share which decreases investment and thereby decreases overall productivity.

Employment thus decreases.

This results in lower wages and the cycle begins anew.

Goodwin's model has been called a model of classical economics because of the assumptions underlying the model. Because of the analogy to predator prey, it has also been described as a class struggle model between the worker's income and the capitalist's income. Whatever description one chooses, it does predict periodic orbits. However the economic cycles it predicts are not limited to the range of the variables $(0 \leq e, \Theta \leq 1)$. They are equivalent to the population cycles predicted by the Lotka-Volterra model: closed cycles surrounding a center, hence not structurally stable[5].

---

[5]It was proven in 1966 that structural stability is generically false in systems with more

Goodwin's original model has naturally been extended in many directions by macro-economists with a major theme being to make the modeling assumptions sharper so that the cycles produced are a structurally stable limit cycle, Gabisch and Lorenz [1989]. One general approach is to introduce more nonlinearities in the system by using a nonlinear wage function, rather than its linear approximation, and then to seek a stable limit cycle by appealing to the Poincaré-Bendixon Theorem. Another approach to modeling market cycles is to use a piecewise linear, regime switching, model for the wage function.

The ultimate goal of business cycle models is to inform decisions on counter-cyclical policies and smooth the business cycle, the financial (credit) cycle and protect economies from systemic vulnerabilities. Model accuracy is not (yet) high and globalization means that national economies are linked worldwide. It seems like predictive models will need a much better understanding (developed through analysis of simplified ones), complex algorithms and extensive computational resources. Model accuracy is an important and formidable goal.

9.6 References for Chapter 9

M. Bronfenbrenner, Is the business cycles obsolete?, John Wiley and Sons, New York,1969.

C. Corduneanu, Almost periodic functions, Wiley, 1968.

C. Chirella, The Elements of a Nonlinear Theory of Economic Dynamics, Lecture notes in economics and mathematical systems, volume 343, Springer, Berlin, 1990.

C.W. Clark, Mathematical Bioeconomics, Wiley, New York, 1976.

T. Ferri and E. Greenberg, The labor market and business cycle theories, Lecture notes in economics and mathematical systems, volume 325, Springer, Berlin, 1989.

G. Gabisch and H.-W. Lorenz, business cycle Theory, second edition, Springer, Berlin, 1989.

J. Goldstein, Prosperity and war in the Modern Age, Yale Univ. Press , New Haven, 1988.

R.M. Goodwin, The nonlinear accelerator and the persistence of business cycles, Econometrica, volume 19 (1951) pages 1-17.

R.M. Goodwin, A Growth Cycle, p. 54-58 in: Socialism, Capitalism and Economic Growth (C. H. Feinstein Editor) Cambridge U. Press, Cambridge, 1967.

R.M. Goodwin, W. Kruger and A. Vercilli, Nonlinear models of fluctuating growth, Springer, Berlin, 1984.

---

than 3 equations:

S. Smale, Structurally stable systems are not dense, Am. J Math., 87(1966) 491-496.

Thus there seems to be 2 possibilities. Physically meaningful systems may be important but scarce because they are structurally stable or structural stability is theoretically attractive but a dead end for understanding models.

P. Hayes, Mathematical models in the social and managerial sciences, Wiley, New York , 1975.

H. Lorenz, Goodwin's nonlinear accelerator, Journal of Economics, Volume 47 (1987), pages 413-418.

C. Perez, Technological Revolutions and Financial Capital, Edward Elgar Publishing, Chentenham, 2002.

9.7   Exercises for Chapter 9

1. Pick the parameters in Goodwin's model to be digits from a phone number and sketch the resulting phase portrait.

2. Show analytically that Goodwin's model is conservative if $\frac{1}{r} > (d + n)$. Analyze conservation if $\frac{1}{r} < (d + n)$.

3. Consider a business cycle as modeled in this chapter. What occurs to the cyclic behavior if the worker productivity is suddenly increased/decreased? Does the observed response depend upon what exact point in the cycle productivity changes?

4. How would the model we consider in this chapter be modified if either/both the number of workers or/and the labor productivity is assumed to grow logistically? Find a suitably modified system and plot a representative phase portrait.

5. Business cycles come into and out of fashion cyclically. Do a search for articles on business cycles in newspapers to see if business cycle theory is currently in fashion. Try searching for titles containing "business cycle".

6. If the equilibrium point is taken to be an average over one period of the dynamic behavior, what parameters influence the wage share and the employment ratio?

7. A simple model of an economy: A nation's economy is described by 3 variables

$$I(t) \quad = \quad \text{national-income,}$$

$$C(t) \quad = \quad \text{Consumer spending,}$$

$$G(t) \quad = \quad \text{Govt-spending.}$$

Suppose $G(t)$ is known.

a. What assumptions lead to the following model:

$$I' = I - \alpha C,$$

$$C' = \beta(I - C - G),$$
$$where$$
$$\alpha \geq 1, \beta \geq 1.$$

b. If $G(t)$ is constant, analyze the model.

c. If $G(t)$ grows linearly, analyze the model.

d. If $G(t) = A + BI^2$ analyze the model.

8. Each assumption presented for Goodwin's model includes a suggestion of a more realistic assumption. Pick one and analyze the model. Describe the difference between model prediction in the idealized and slightly less idealized models. For example, analyze the model if either population or productivity grows logistically.

9. In Goodwin's model suppose additional consumption beyond the wage share subtracts from investment and obeys a "consumption accelerator" law meaning

$$C(t) = AE(t) + BE'(t).$$

Develop the model under this new assumption and analyze its predictions.

Bibliography

[AM79]   R.M. Anderson and R.M. May, Population biology of infectious diseases, part 1, Nature 280 (1979) 361- 367, part 2, Nature 280 (1979) 455-461.

[AC49]   A.A. Andronov and C.E. Chaikin, Theory of Oscillations, Princeton University Press, Princeton, 1949.

[A73]   V.I. Arnold, Ordinary Differential Equations, Massachusetts Institute of Technology Press, Cambridge, Massachusetts, 1973.

[AP99]   D. Arnold and J.C. Polking, Ordinary Differential Equations using MATLAB, Prentice Hall, 1999.

[B10]   N. Bacaër, A short history of mathematical population dynamics, Springer, Berlin, 2010.

[B75]   N.T.J. Bailey, The mathematical theory of infectious diseases, London, Griffin, 1975.

[BF97]   L. Beierora and J.W. Forrester, Generic structures: Overshoot and Collapse, MIT, press, 1997.

[BT98]   J.A. Brander and M.S. Taylor, The simple economics of Easter Island: A Ricardo-Malthus model of renewal resource use, The Am. Econ. Review 88 (1998) 119-138.

[B69]   M. Bronfenbrenner, Is the business cycles obsolete?, John Wiley and Sons, New York,1969.

[CC96]   The carrying capacity briefing book, the Carrying Capacity Network, Washington, D.C., volumes 1 and 2, 1996.

[C90]   C. Chirella, The Elements of a Nonlinear Theory of Economic Dynamics, Lecture notes in economics and mathematical systems, volume 343, Springer, Berlin, 1990.

[C76]   C.W. Clark, Mathematical Bioeconomics, Wiley, New York, 1976.

[C70]   D.O. Cogwill, The use of the logistics curve and the transition model in developing nations, in: Studies in demography, (Editors: Boes, Desai and Hain) University of North Carolina press, Chapel Hill, 1970.

[C95]   J.E. Cohen, How many people can the earth support?, W.W. Norton and Co., New York, 1995.

[C68]   C. Corduneanu, Almost periodic functions, Wiley, 1968.

[C77]   J. Cronin, Some mathematics of biological oscillations, SIAM Review, volume 19 (1977), 100-138.

[D54]   U. D'Ancona, The struggle for existence, Leiden, Brill, 1954.

[D38]   J. Davidson, On the Growth of the Sheep Population in Tasmania, Trans. R. Soc. S. Australia 62(1938): 342–346

[DW00]   M.L. Deaton and J.J. Winebrake, Dynamic modeling of environmental systems, Springer, NY, 2000.

[D72]   D. Defoe, A journal of the plague year, London, Oxford University Press, 1972.

[D05]   J. Diamond, Collapse: How societies choose to fail or succeed, Viking, NY, 2005.

[D67]   K. Dietz, Epidemics and rumors: a survey, Journal of the Royal Statistical Society, 130 (1967), 505-528.

[E68]   P.R. Ehrlich, The population bomb, Valentine books, New York, 1968.

[E73]   T.C. Emmell., An introduction to ecology and population biology, New York, Norton, 1973.

[FG89]   T. Ferri and E. Greenberg, The labor market and business cycle theories, Lecture notes in economics and mathematical systems, volume 325, Springer, Berlin, 1989.

[GL89]   G. Gabisch and H.-W. Lorenz, Business Cycle Theory, second edition, Springer, Berlin, 1989.

[G34]   G.F. Gause, The struggle for existence, Williams and Wilkins, Baltimore, 1934.

[G73]   M.E. Gilpin, Do hares eat lynx?, American Nat. v.107 (1973) 727-730.

[G88]   J. Goldstein, Prosperity and war in the Modern Age, Yale Univ. Press , New Haven, 1988.

[G51]   R.M. Goodwin, The nonlinear accelerator and the persistence of business cycles, Econometrica, volume 19 (1951) pages 1-17.

[G67]   R.M. Goodwin, A Growth Cycle, p. 54-58 in: Socialism, Capitalism and Economic Growth (C. H. Feinstein Editor) Cambridge U. Press, Cambridge, 1967.

[GKV84] R.M. Goodwin, W. Kruger and A. Vercilli, Nonlinear models of fluctuating growth, Springer, Berlin, 1984.

[H75]   P. Hayes, Mathematical models in the social and managerial sciences, Wiley, New York , 1975.

[H76]   H.W. Hethcote, Qualitative analyses of communicable disease models, Math. Biosciences 28 (1976).

[HS74]   M.W. Hirsch and S. Smale, Differential equations, dynamical systems and linear algebra, Academic Press, London,

[H59]   C.S. Holling, The components of predation, Canadian Entomol. 91(1959) 293-320.

[H95]   F. Hoppensteadt, Getting started in mathematical biology, Notices of the AMS, Sept. (1995)969-975.

[H06]   A. Householder, The theory of matrices in numerical analysis, Dover, 2006.

[HW92]   J.H. Hubbard and B.H. West, MacMath 9.0, A dynamical Systems Package for the MacIntosh, Springer, Berlin, 1992.

[H45]   M.K. Hubbert, Energy from fossil fuels, Science 109(1945)103-109.

[JS77]   D.W. Jordan, and B. Smith, Nonlinear Ordinary Differential Equations, Clarendon Press, Oxford 1977.

[K03]   E. Kalnay, Atmospheric Modeling, Data Assimilation and Predictability, Cambridge Univ. Press, 2003.

[KM27]   W.D. Kermack and A.G. McKendrick, A contribution to the mathematical theory of epidemics, Journal of the Royal Statistical Society, 115 (1927), 700-721.

[K96]   D. Kirschner, Using mathematics to understand HIV immune dynamics, Notices of the AMS, February, (1996) 191-202.

[K72]   C.J. Krebs, Ecology: The Experimental Analysis of Distribution and Abundance, Harper and Row, New York, 1972.

[K89]   H. Kocak, Differential and Difference Equations through Computer Experiments, Springer-Verlag, Berlin, 1989.

[L56]   F.W. Lancaster, Mathematics in Warfare, p. 2138-2157 in: The World of Mathematics, J.R. Newman editor. Simon & Schuster, 1956.

[LF06]   P.E. Lekone and B.F. Finkenstadt, Statistical inference in a stochastic SEIR model with control interaction: Ebola as a case study, Biometrics 62(2006) 1170-1177.

[L87]   H. Lorenz, Goodwin's nonlinear accelerator, Journal of Economics, Volume 47 (1987), pages 413-418.

[L25]   A.J. Lotka, Elements of physical biology, Williams and Wilkins, Baltimore, 1925.

[L20]   A.J. Lotka, A new conception of the universe, Harpers monthly magazine, March 1920, pages 477 - 487.

[L01]   S. Lynch, Dynamical Systems with Applications using Maple, Birkhauser, 2001.

[MM76]   J.E. Marsden and M. McCracken, The Hopf bifurcation and its applications, Applied Mathematical Sciences, Volume 19, Springer Verlag, New York, 1976.

[M74]   R. May, Stability and Complexity in Model Ecosystems, Princeton University Press, Princeton New Jersey, 1974.

[MA87]   R. May and R.M. Anderson, Transmission dynamics of HIV infection, Nature, 326 (1987) , 137.

[M93]   J.D. Murray, Mathematical Biology, Springer, Berlin,1993.

[M98]   I. Mahon, Easter Island: The economics of population dynamics and sustainable development in the Pacific context, 113-119 in: Easter Island in Pacific context: South Seas Symposium (C.M. Stevenson et al., editors), Easter Island Foundation, Los Osos, 1998.

[M70]   T.R. Malthus, An essay on the principle of population, and a summary view of the principle of population, collected and edited by A. Few, Penguin, Baltimore, 1970.

[M37]   D.A. MacLulich, Fluctuations in the numbers of the varying hare (Lepus americanus). University of Toronto Studies Biological Series 43. University of Toronto Press, Toronto, 1937.

[M33]   J.R. Miner, Pierre-Francois Verhulst, the discoverer of the logistic curve, Human Biology, 5 (1933), 673-685.

[O53]   E.P. Odum, Fundamentals of Ecology, W.B. Saunders, London, 1953.

[O78]   M. Olinick, An Introduction to Mathematical Models in the Social and Life Sciences, Addison-Wesley, Reading, Mass., 1978.

[PR20]   R. Pearl and L. Reed, On the rate of growth of the United States population since 1790 and its mathematical representation, Proceedings National Academy of Science, 6 (1920) 275-288.

[P41]   R. Pearl, Some biological considerations about war, American Journal of Sociology, Volume 46 (1941), pages 487 - 503.

[P02]   C. Perez, Technological Revolutions and Financial Capital, Edward Elgar Publishing, Chentenham, 2002.

[P69]   E.C. Pielou, An Introduction to Mathematical Ecology, Wiley, New York, 1969.

[P73]   H. Pollard, Mathematical models of the growth of human populations, Cambridge University Press, Cambridge 1973.

[R56]   L.F. Richardson, Mathematics of War and Foreign Politics, pages 1240-1253 in: The World of Mathematics, vol. IV (J. R. Newman, editor) , Simon and Schuster, New York, 1956.

[R39]   L.F. Richardson, Generalized Foreign Policy, British J. of Psychological Monographs, Supplement 23, 1939.

[R60]   L.F. Richardson, Arms and Insecurity: A Mathematical Study of the Causes and Origins of War, Boxwood Press, Pittsburgh, 1960.

[RM63] M.L. Rosenzweig and R.K. McArthur, Graphical representation and stability conditions of predator-prey interactions, Am. Natur. 47(1963) 209-223.

[R11]   R. Ross, The prevention of malaria, second edition, London, Murray, 1911.

[S68]   T.T. Saaty, Mathematical Models of Arms Control and Disarmament, Wiley, New York, 1968.

[S57]   H.A. Simon, Models of Man: Social and Rational, Wiley, 1957.

[S61]   L.B. Slobodkin, Growth and regulation in animal populations, New York, Holt, Rheinehart and Winston, 1961.

[S75]   L.B. Slobodkin, Comments from a biologist to a mathematician, pages 318-329 in: Ecosystem analysis and prediction, (S. A. Levine, editor), SIAM, Philadelphia, 1975.

[S66]   S. Smale, Structurally stable systems are not dense, Am. J. Math., 87(1966) 491-496.

[S29]   H.E. Soper, Interpretation of periodicity in disease prevention, Journal of the Royal Statistical Society, 92 (1929), 34-73.

[SS02] L.G. Stanley and D.L. Stewart, Design sensitivity analysis: computational issues of sensitivity equation methods, SIAM, 2002.

[S74]   V. Steffler, Long term forecasting and the problems of large scale wars, Futures 6(1974) 302-308.

[T68]   J.T. Tanner, The stability and intrinsic growth rate of prey and preda-
          tor populations, Ecology, 56 (1968), 855-867.1974.

[T88]   J. Tainter, The collapse of complex societies, Cambridge U. Press,
          Cambridge, 1988.

[V31]   V. Volterra, Lecons sur la Theorie Mathematique de la Lutte pour la
          Vie, Gauthier-Villars, Paris, 1931

[W74]   P. Waltman, Deterministic threshold models in the theory of epi-
          demics, New York, Springer, 1974.

Index

www.ingramcontent.com/pod-product-compliance
Lightning Source LLC
Chambersburg PA
CBHW082035190526
45165CB00021B/3327